WATCHING GIANTS

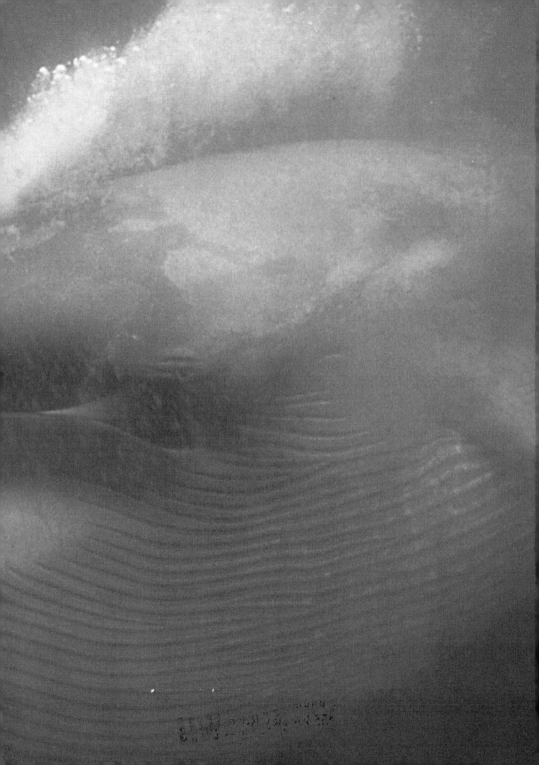

WATCHING

GIANTS

THE SECRET LIVES OF WHALES

ELIN KELSEY

With Photographs by Doc White
Additional Photographs by François Gohier

UNIVERSITY OF CALIFORNIA PRESS   Berkeley Los Angeles London

The publisher gratefully acknowledges the generous support of the General
Endowment Fund of the University of California Press Foundation.

University of California Press, one of the most distinguished university presses
in the United States, enriches lives around the world by advancing scholarship
in the humanities, social sciences, and natural sciences. Its activities are supported by the UC Press Foundation and by philanthropic contributions from
individuals and institutions. For more information, visit www.ucpress.edu.

University of California Press
Berkeley and Los Angeles, California

University of California Press, Ltd.
London, England

Library of Congress Cataloging-in-Publication Data

Kelsey, Elin.
  Watching giants : the secret lives of whales / Elin Kelsey ; with photographs
by Doc White; additional photographs by François Gohier.
      p.   cm.
  Includes bibliographical references.
  ISBN 978-0-520-26158-7 (pbk. : alk. paper)
  1. Whales.   2. Whales—Mexico—California, Gulf of.   I. Title.

QL737.C4K434 2008
599.509164'1—dc22   2008007782

Manufactured in Canada

18 17 16 15 14 13 12 11 10 09
10 9 8 7 6 5 4 3 2 1

The paper used in this publication meets the minimum requirements
of ANSI/NISO z39.48—1992 (R 1997) (Permanence of Paper).

For the inspiring mothers in my family:
Grace Kelsey, Janice Harper, Karen Kelsey,
Alison Kelsey, Pamela Johnson,
and, of course, Kelly

The female whale or dolphin, in her role as a mother,
is the linchpin of cetacean population
and behavioural biology.

Hal Whitehead and Janet Mann,
Cetacean Societies

I've got a good mother,
And her voice is what keeps me here.

Jann Arden, *Good Mother*

CONTENTS

Photographs follow page 82

ACKNOWLEDGMENTS

"All men are islands," says Will in the first few lines of *About a Boy,* Nick Hornby's wonderful screenplay about fatherhood, responsibility, and the struggle to grow up. Most women, I would wager, harbor no such delusion. The number of people involved in the creation of this book could fill a small island state. First and foremost, I wish to thank the scientists and conservationists whose thoughts and feelings are the backbone of this book. These individuals generously made time to share their experiences, no matter how hectic their workloads and travel schedules or how probing my questions. They include Alejandro Acevedo-Gutiérrez of Western Washington University; Bernie Tershy and Don Croll of the University of California, Santa Cruz; Bob Brownell and Jay Barlow of the Southwest Fisheries Science Center; Boris Worm, Hal Whitehead, and Shane Gero of Dalhousie University; Bruce Mate of Oregon State University; Christopher Clark of Cornell University; Craig Smith of the University of Hawaii at Manoa; Diane Gendron of the Interdisciplinary Center of Marine Sciences (CICIMAR); Enric Sala of the Scripps Institution of Oceanography; Exequiel Ezcurra of the San Diego Natural History Museum; Fred Sharpe of the Alaska Whale Foundation; Gabriela Anaya Reyna of Niparajá; Jorge Urban and Mercedes Eugenia Guerrero Ruiz of the Autonomous University of Baja California Sur (UABCS); Juan Pablo Gallo Reynoso and Lloyd Findley of Mexico's Research Center in Feeding and Development (Guaymas Unit); Liz Alter and William Gilly of Hopkins Marine Station of Stanford University; Lori Marino of Emory University; Mike McGittigen of SeaWatch; Nathalie Jaquet of the

Provincetown Center for Coastal Studies; Phil Clapham of the Alaska Fisheries Science Center; Sandy Lanham of Environmental Flying Services; Alejandro Robles of Noroeste Sustentable; Lance Barrett-Lennard of the Vancouver Aquarium Marine Science Centre; and Steve Webster of the Monterey Bay Aquarium. It fills me with hope to know that so many talented, committed, and compassionate individuals are working on behalf of whales, science, and the ocean.

When it comes to conservation heroes, I would especially like to honor Ernesto Bolado Martínez and Maria de los Angeles Carvajal for their inspiring dedication to the Gulf of California and for their extraordinary generosity and willingness to look beyond the confines of bureaucracy to create a writer-in-residency role for me with Conservation International Mexico that mutually supported their program and the background research for this book.

This book would not be the book it is without the support and wisdom of Jenny Wapner, my editor at UC Press. There is nothing as wonderful for a writer than an editor who is passionately supportive of trying something new and who gracefully protects the idea as it passes through the tricky stages of coming into being. Sincere thanks also go to Tierney Thys, Fred Sharpe, and Peter Moyle for their rich and thorough scientific reviewer's comments, and to Ellen (Dale) Russell and Paul Mishler for reading early drafts. Who would have thought an economist and a labor historian would be the ideal people to help me find my woman's voice in natural history writing?

It is far easier to write when you know your children are off having a great time. Thank you to Maria del Lourdes Zavala Reynozo (MaryLu) and her loving family, Jose Luis, Luis, MaryFer, and Roscio, for making Kip and Esmé (and Andy and me) feel so much a part of your life in La Paz, Mexico. Huge hugs as well to Gai Quigley, Laura Friesen, and Fiona Harper for all the stories and songs and bike rides and craft projects and shifts at the preschool and kindergarten back home in Pacific Grove, California.

A quiet space to write is one of the holy grails of a writer's life. I remember hearing Andre Dubus III describe how he wrote *House of Sand and Fog* while sitting in his truck in the middle of a graveyard. I too am lucky enough to live in a tiny, child-filled home and am deeply grateful to Roxane Buck-Ezcurra, Judy Long, Sarah Paff, and Debbi Neel for leaving their keys under their mats and coffee on their counters when I really needed a peaceful place to focus on the page. I also wish to thank Peter Barnes and Leyna Berstein for creating the Mesa Refuge and for giving me the gift of two solid weeks in which to write and revel in the beauty of Pt. Reyes National Seashore.

Sincere thanks go to Paul Fleischman for the perfectly timed letter about authenticity and failure and for being a true mentor and friend. The final words go to the great loves of my life, Kip, Esmé, and my husband, Andy Johnson, who finds endless ways to love, support, and surprise me. (How rare to find a marine mammal expert, talented copyeditor, and world's best dad all in one man.) I am lucky, lucky, lucky.

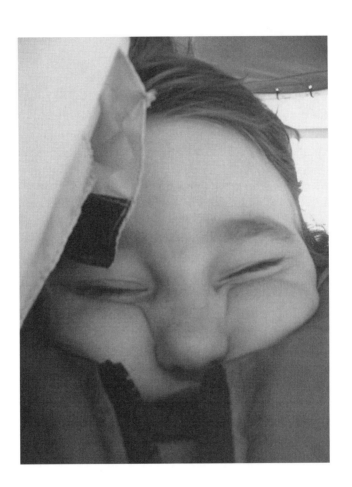

This is a picture of my daughter, Esmé, whale watching. She is three and a half, and as her six-year-old brother, Kip, will tell you, if you want to get Esmé to go to sleep, all you need to do is turn on the *Goodnight Moon* tape or put her on a whale watching boat.

The waters surrounding Mexico's Baja Peninsula are the best in the world for seeing the greatest diversity of whales. Little did I know, when I began scheming to bring my children with me to the Gulf of California to inter-view whale researchers for this book, that what Esmé would really get out of the deal was plenty of rest. She slept through the gray whale calving lagoons along the Pacific Coast. She slept through the blue whales and a thousand common dolphins off Loreto in the southern gulf. She even managed to sleep on a whale watching boat that turned out to be a booze cruise—complete with free drinks and the Bee Gees blaring from the stereo as it roamed for humpback whales off the southern tip of Cabo San Lucas.

There is no mention, in the hundreds of glossy brochures I've seen, of the sleep-inducing powers of whale watching. On the contrary, whale watching advertisements promise "ultimate" and "once-in-a-lifetime" experiences and feature photos of excited people shielding themselves from the ocean spray generated by whales leaping alongside their boats on brilliantly sunny days. Such images have parlayed whale watching into a billion-dollar industry—a jewel in the crown of ecotourism, the fastest-growing segment of the world's largest industry, tourism.[1]

But like the child who speaks the truth in "The Emperor's New Clothes,"

I think Esmé's take on whale watching is pretty honest. After the initial exhilaration of casting off from the dock, trying on life jackets, fiddling with cameras, and eating all the snacks you brought for later, people on a whale watching boat begin to resemble subway commuters on their way home from work. Some chat companionably. Some stare off into space. Bundled in layers to stave off the chill, most of them, like Esmé, inevitably fall asleep.

So why do 10 million of us every year rearrange our vacations and shell out lots of money for an experience more readily and more affordably found on the subway? The answer, of course, is that going whale watching is nothing like taking public transit. When you see the glimpse of a whale's back from fifty yards away, you don't see it as the tiny gray speck, in a sea of gray water, that will eventually show up in frame after frame of your holiday snapshots. No, in your mind's eye, you see the extraordinary things that you have been told whales do. You have to know the stories to get the thrill.

Years ago I took a whale watching trip off Baffin Island in the Canadian High Arctic. I remember vividly the sensation of life all around me as I sped across an ice floe on the back of a dogsled in pursuit of belugas and narwhals. It was only after I was left standing on the ice edge, staring into the dark open water on my own, that I experienced the Arctic as a cold, empty place. My sensation of its richness was almost entirely vicarious, felt through the stories of the Inuit guide who showed me ringed seal holes and the downy feathers of eider ducks as we traveled. In his simple act of leaving to get supplies for tea, he swept the life from this foreign environment, as easily as one brushes crumbs from a table.

This book immerses you in the foreign world of whales. It is filled with the voices of more than twenty-five of today's leading talents in whale research and marine ecology, people like Hal Whitehead, professor of biology at Dalhousie University, who brings his gifts as a mathematician to the study of sperm whale culture and Sandy Lanham, an eco-pilot who flies her 1958 Cessna plane with Mexican researchers to help them track endangered blue whales across the Gulf of California. Each is an insider who can see things that you and I can't—not just "can't" because we're unlikely to have the time, resources, ability, expertise, or desire to spend hundreds of hours at sea trying to observe the lives of animals that spend 90 percent of their time underwater and that frequently move around the globe far from shore and in rough weather, but also "can't" in a very real sense. It's an issue of scale. The situation is much like that of the classic children's tales, *The Mouse and the Motorcycle* and *The Borrowers,* stories based on the premise that we share our homes with individuals so small they escape our notice. In the case of whales, we share the planet with individuals and populations that operate on a scale too *large* for us to comprehend using just our human senses of sight, hearing, taste, touch, or smell.

Humpback whales living off the coast of Norway offer an interesting case in point. Spot one in the ocean, and unless it's with a calf, it's likely to be alone. It took the end of the cold war, and the navy's decision to permit civilian scientists to use their antisubmarine naval listening stations, for Christopher Clark, director of the Bioacoustics Research Program at Cornell University, to be able to track whales using computer images of individual whale's voices, and in doing so to reveal a different picture. "I can watch cohorts of humpback whales off the coast of Norway, singers, that are separated by ten or twenty or thirty miles," he tells me. "These whales move as a collection—an acoustic herd—over days and weeks. If you and I were whale watching, we would be totally unaware that this is going on."

What's intriguing about all this is that even expert researchers find it challenging to study cetaceans. Many populations of whales, dolphins, and porpoises spend time offshore, sometimes in rough, stormy waters that are difficult, if not impossible, to access. They move according to the distribution of their food. They move with the seasons. They change their movements in response to El Niños and other environmental events. They travel great distances at fast speeds. "Most of what people have said whales are doing," Bruce Mate, director of the Oregon State University Marine Mammal Program states, "is based on what they see whales doing at the surface, which is mostly breathing!"

Given these realities, it's easy to appreciate why Christopher confesses to having envied scientists who study animals on dry land. Recently he joined a colleague, Andrea Turkalo, in Central Africa to explore how forest elephants use sounds. He found it extraordinary that she could watch individuals for hours or days at a time. "It was unbelievable," he says. "You could really see how Andrea could learn the personalities of these different elephants, because you can watch them behave. Aunts, mothers, sisters. . . . I couldn't help but think that this is what's going on underwater that we never get to witness."

In the town of Pacific Grove, California, where I live, there's a life-size cement replica of a baby gray whale outside the local natural history museum. It's become a rite of passage for those under six in the neighborhood to try to shimmy up the tail and walk across the back before "blowing" themselves from the blowhole into the safety of the sand below. Only a few rare people experience living whales at a scale this intimate. Doc White, a marine mammal photographer whose photos are included in this book, is one of them. He and his wife, the artist Ceci Devereaux, have traveled the world in pursuit of opportunities to don snorkels and masks and slip into the water with whales. Ceci tells the story of dropping Doc off the starboard side of their small inflatable boat close to a blue whale in Magdalena Bay (Bahía Magdalena) along the Pacific Coast of Baja California. "A moment later the whale surfaces on the port side, and I'm touching a blue whale!" she says. "Doc's under the boat, and I'm thinking he's got the greatest picture *ever*. But then he comes up and asks me, 'Where's the whale?' The krill was so thick he missed an eighty-foot blue whale swimming on top of him." Whales can prove elusive prey, even to those in close proximity and possessing undeniable expertise.

Like in the parable of the blind men and the elephant, whales are too big, too foreign, too complex, and too diverse for any one person, researcher, or whale enthusiast to understand. You need to look across species, across different research approaches, across time to "see" the extraordinary lives of whales. Maybe because they are so challenging to observe, whales attract innovative thinkers. In the past decade or so, a number of whale researchers have pioneered new uses of technology, some of it borrowed from the cell phone industry and genetics labs, to study whales and their environments in both more intimate and more massive scales. FedEx packages of whale skin and blubber are flown to genetics labs, where chemists test them for toxicity and geneticists trace family origins. Crittercams strapped to the backs of humpbacks provide a whale's perspective on the way teams of bubble-blowing whales assemble themselves to catch fish. Satellite tags transmitting information about where whales travel beam up to space and back to the Internet, where we can track the whales' movements across our BlackBerries while checking our email.

What's become clear is that complex societies of whales are living in the ocean. And they differ from one another in fascinating ways. A blue whale or a fin whale has a totally different ecology and social system than a humpback whale. Mexico's Gulf of California provides the ideal location from which to dive into the subject of whale cultures. Thanks to its unique ecology, geology, and oceanographic features, it is an enormously rich sea. It spans an area the size of Ecuador—about 109,000 square miles—and is home to nearly 900 species of fish and a third of the world's species of whales, dolphins, and porpoises. Jacques Cousteau called it "the Aquarium of the World." Though the research described in this book hails from many parts of the planet, each of the whale and dolphin species discussed may be found in this miraculous sea.

The structure of this book mirrors the complexity of its subject. Each chapter is a separate essay that can be enjoyed on its own. Yet reading all the chapters in their entirety will give you the richest understanding of whale and dolphin cultures and the intriguing ways scientists attempt to comprehend them. From the meaning of menopause in sperm whales to the role of teenage killer whale babysitters to the discovery of bone-eating zombie worms found only on the sinking bodies of dead whales, this book examines cetacean cultures in new and compelling ways.

You might also notice that it is written from my perspective as a woman and a mother. I've noticed a tendency for popular science and natural history books to tell tales of grand adventure in which a writer, often a man, leaves his everyday life behind to immerse himself in the jungles of Borneo or the icy mountains of Antarctica. Such accounts are thoroughly engaging, but they sometimes leave me wondering about the lives of those left behind. What would natural history books look like if mothers and children lived the adventures?

Whales, I have discovered, are extremely dedicated mothers. They have to be. There are no safe places to leave one's baby to rest or feed in the ocean. Whale motherhood is a 24/7 enterprise. And for some species, it is a lifetime role. Male killer whales, for example, stay with their mothers their entire lives, lives that may span several decades. The Gulf of California is, in essence, the largest known nursery for the biggest babies on the planet.

It seems only fitting, therefore, to explore the science of whale motherhood, social relationships, and cultures against the backdrop of the Gulf of California, through my combined identity as writer and mother of two young children. I was fortunate enough to have been granted a writer-in-residency position that made such a family experience possible. In the course of writing this book, Kip and Esmé attended school in Mexico, toured research labs, collected whale poop, played countless games of tag with the children of whale researchers, and went whale watching—a lot.

Clearly some women will try anything to get their kids to sleep.

1

Extreme Motherhood

The cartoon gracing the cover of the May 2006 issue of *The New Yorker* depicts an exhausted mother pushing an immense stroller overflowing with kid paraphernalia. The load is reminiscent of the one Max, the Grinch's dog, had to drag away from Whoville. My friend Heidi's jogging stroller makes such images look spartan. The back pouch strains under the weight of juice boxes, water bottles, and snacks of the fruit, cheese, cracker, carrot, snap pea, and cookie variety. An umbrella, flip-flops, wet wipes, and sunscreen struggle free of pockets on the sides. The lower net contains a noisy clatter of pails, shovels, and toys. Packages to be mailed, library books, and school projects perch on top. Most mornings, the whole contraption comes complete with a ninety-pound Portuguese water dog tied to the handle, and two kids. It's a double.

Heidi's stroller is a symbol of twenty-first-century mothers on the go, mothers who take their kids with them everywhere, and thus, come well prepared. I cannot imagine a single eventuality for which Heidi would not be ready, given the supplies at hand. Perhaps that is why I am so struck by the contrast. While Heidi moves her enormous load of necessities along the recreation trail overlooking the coastline of Monterey Bay, mother gray whales *(Eschrichtius robustus)* are leading their calves so close to the shore that Heidi's little ones can see their blows. These other mothers are embarked on a round-trip journey from Mexico to the Arctic and back again—roughly the equivalent of 640,000 lengths of an Olympic swimming pool. They follow one of the longest migration routes of any mammal on earth, traveling nearly thirteen thousand miles round-trip. They travel day and night, covering an average of ninety miles every twenty-four hours. And they take nothing with them.

Once she's sexually mature, a gray whale gives birth to a calf every other year. Which means that in every twenty-four-month cycle, she's pregnant or lactating 80 percent of the time. Despite the impressive caloric demand, she virtually fasts (or eats very little) from the time she begins her migration south until she returns to the northern feeding grounds six months later.[1] So a pregnant female arriving in the Baja Mexico lagoons in the first part of January won't get back to food until May. She will fast that whole time and deliver a one-ton baby and feed it with milk that is 50 percent fat. The energy demands are staggering.

Mercedes Eugenia Guerrero Ruiz is a whale researcher at the Autonomous University of Baja California Sur (UABCS), one of Latin America's leading marine biology institutes. She's also the mother of a toddler. When we meet on a clear March morning in a café along the Malecón, a wide welcoming seashore promenade rimming the turquoise waters of La Paz Bay, I am running late and thankful that the woman sitting across from me understands the challenges of trying to marry professional schedules with a three-year-old's concept of time. It's hard not to think about children when you're in La Paz. The streets of the beautiful capital city of Baja California Sur are lined with dozens of language schools and kindergartens, each aflutter with neatly dressed children in brightly colored uniforms. A quarter of a million people live in this thriving commercial, cultural, and political center, and thanks to one of the highest standards of living in the country, the population is growing fast.

Mercedes and I spend a few moments getting acquainted and laughing over the impossibility of making the tiny feet in those tiny shoes *please go faster* when we're late, when the talk shifts to whale mothers. "Just imagine," she says, reflecting on how vulnerable she sometimes feels when her daughter falls too far behind. "It must be the same for the whales. They have to be aware of predators, but if they are being harassed and they have a calf, they can't move fast."

The ocean is not an easy place to be a baby—or a single mother. Cetaceans are the only group of mammals to evolve in a habitat where there are so few places to hide. Individuals of some dolphin or whale species try to escape particular predators by swimming close to shore or diving deep, but for most cetaceans, the only refuge from predators is found in group living. Gray whale mothers gather in the thousands along the Baja Peninsula's four major calving lagoons—Laguna Guerrero Negro, Laguna Ojo de Liebre (formerly known as Scammon's Lagoon), Laguna San Ignacio, and Bahía de Magdalena—to give birth, presumably because these bays offer protection from bad weather and from sharks and other large predators.

Few people have witnessed the birth of any whale species in the wild. Sandy Lanham is one of the fortunate souls who has. Sandy is the pilot for and the founder of Environmental Flying Services, a nonprofit organization that supports research flights for biologists in Mexico. In 2001 the MacArthur Foundation granted her one of its coveted genius grants—$500,000 with no strings attached—in recognition of her extraordinary originality and dedication to science and conservation. She frequently flies researchers doing aerial surveys over Baja's gray whale calving lagoons, sometimes seeing as many as 700 gray whales in a day. The day she saw the whale birth, she was flying scientific transects over Laguna Ojo de Liebre, a UNESCO World Heritage Site and home to more than half the gray whale births in Baja. "It took a minute to figure out that this was a whale giving birth," she says, describing the sight of a little fluke poking beneath a mother's belly in the very shallow water at the eastern end of the lagoon. But rather than grab for her camera, she instinctively turned the airplane away as violently as she could. "I wondered how the researchers, two men, would react to that. But we all agreed we had no business being there."

All Sandy's actions demonstrate her sensitivity to animals and her appreciation for the tenderness of whale mothers. "I've seen instances with gray whales, when the airplane would scare them," she tells me. "The calf would hurry back to her mother, and the mother would stretch her pectoral flipper to just gently brush the calf." Every case moves her. "We're always cautioned against reading emotion into wildlife; there's a reluctance to ascribe feelings to whales," she says. "But the thing is, what other binding factor is there except for emotions? What else is going to ensure that a mother takes care of her young?"

Gray whale mothers and their calves stay in the lagoons for two months so that the babies can grow strong enough to migrate back. By March or April, small groups of four or five mother-calf pairs leave the safety of the lagoons and begin the treacherous trip northward. There's a good chance the whales you can see from the California coastline each spring are mothers and calves. They hug the coastline and take advantage of places like Point Conception, the biggest point that sticks out in Southern California, to seek temporary shelter from rough weather and to find a protected place for the calf to suckle.

A gray whale calf depends entirely on its mother. She is the one who will nurse it, show it how to navigate one of the world's busiest coastlines to the northern feeding grounds, and eventually, how to feed on its own. The trouble is, from a gray whale's perspective, there are other whale mothers in the ocean teaching their calves too—killer whale *(Orcinus orca)* mothers.

The bucolic shore of Monterey Bay where Heidi and her little ones stroll has a more sinister nickname in the whale research community. It's called "ambush alley." Monterey Bay is a prime location for mother killer whales to teach their calves how to feed on gray whale calves. Every April and May, young transient, or mammal-eating, killer whales arrive here to take advantage of this excellent food source. Researchers sometimes compare the activity to a mother cat teaching her kittens how to hunt mice. If you examine any of the pictures of killer whales attacking gray whales in Monterey Bay, you'll see that it's mainly the young animals who try to separate the calf from the mother. The adult female killer whales aim toward the periphery, acting as backup.

The playful cat comparison belies the full horror of the experience. According to Nancy Black, a killer whale biologist at Monterey Bay Whale Watch who has witnessed this astonishing predation event several times, six hours may pass from the time a group of killer whale juveniles and calves give chase to the moment when a gray whale calf is killed. They ram and bite the gray whale calf in their efforts to drag it from its mother. During this period the mother and calf will try to dash for the safety of shallow water. In her desperate attempts to shield the calf from the intense onslaught, the mother repeatedly tries to roll the calf onto her belly. Eventually the calf, too exhausted to continue, is driven from the mother and swiftly drowns. Sometimes the force exerted by the killer whales is enough to decapitate the calf.[2]

It has become a common practice in recent years to try to impress young teens with the demands of motherhood by asking them to strap on a baby carrier and look after a ten-pound bag of flour for a week or to provide round-the-clock care to an egg. What activity, I wonder, could possibly give one a sense of the extreme demands facing a gray whale mother?

2

A Sea of Milk

Wearing your baby is all the fashion these days. Kate Hudson does it. Gwen Stefani does it. Brad Pitt does it. BabyBjörns, Moby Wraps, slings—products that let you hold your baby close—are the most exciting parenting concept to hit the Western world in years, according to baby care experts Dr. William Sears and Martha Sears.[1] Baby wearing is a visible manifestation of a popular parenting style called "attachment parenting." Its primary aim is to help parents and babies connect. I wrapped my children, Kip and Esmé, throughout their first years of life and swear by its ease and comfort for both mother and child. But I must confess to having had moments of claustrophobia. The flip side of having your wonderful baby snuggled close against your chest is the feeling that there is always a baby snuggled close against your chest, or when they reach the toddler stage, wrapped so tightly around your legs that it requires a Houdini move to get to the bathroom for a pee. The physical contact in the first years of mothering feels constant and intense.

Whales and dolphins, obviously, have no arms with which to cradle their babies. They do not hold hands to cross busy shipping lanes. But if country singers wrote lovin' songs about animals instead of people, there'd surely be a lot of lyrics about whale mamas. Whale mothers display a level of attachment to their babies that would humble Dr. Sears. Like us, most cetacean species have long lives and mature late. They reproduce slowly, producing just one calf, in which they invest heavily, at a time. Sperm whale *(Physeter macrocephalus)* mothers, for instance, live into their seventies and suckle each of their babies for up to thirteen *years.* Gray whale mothers are legendary for attacking whaling ships in vain attempts to free their captured young. But if you really want to witness profound examples of maternal attachment, visit the resident killer whales off the coast of Washington and British Columbia. Resident killer whale sons remain mamas' boys forever. They stay with their mothers their entire lives. They're the only male mammals in the world to do so. If that's not fodder for a song, I don't know what is.

The waters surrounding Mexico's Baja Peninsula are the best in the world to see whale mothers and calves. But it would be a mistake to think of it as one big birthing tub. The reason so many whale species, and so many individual mothers, choose to give birth or care for their young calves here is the extraordinary diversity of habitats it offers. "Each whale species comes for a different reason, at different times, in different numbers, to different places," says Jorge Urban, coordinator of the Marine Mammal Research Program of the Autonomous University of Baja California Sur (UABCS). Jorge is one of the gods of marine mammal research in Mexico. He's spoken at more than sixty international meetings and produced dozens of scientific publications about the whales and dolphins of the Gulf of California and the Mexican Pacific. He reminds me of a maître d' as he reels off all the reasons the big names frequent Baja waters: Gray whales and humpback whales are here to breed and calve. Blue whales come to breed, calve, and feed. Fin whales are resident year-round, so they're here for everything. And sperm whales stay with their mothers for several years, so they come here to eat the squid. Jorge estimates that 15,000 to 20,000 gray whales and 3,000 to 4,000 humpbacks visit the Mexican Pacific each year.

All I can think of is the milk. The Gulf of California must be a sea of milk. Whale milk is among the richest on earth. Whipped cream is 30 percent fat. Baleens, such as blue, humpback, and gray whales, kick the fat content even higher, producing milk that is 30 to 53 percent fat. A blue whale mother produces 485 pounds of this superfatty milk each day. She nurses her calf for six months, transferring more than 88,000 pounds of milk to her calf. Imagine the growth chart required to record the calf's weight gain in its first six months: a whopping 37,500 pounds![2]

Weight gain among women, even among pregnant women, can be a loaded topic. Nothing we are capable of doing weightwise, however, comes remotely close to the capacity of pregnant baleen whales to pack on the blubber, and *quickly!* A pregnant blue whale *(Balaenoptera musculus),* weighing 131 tons, deposits more than a third of her body weight in blubber in a single summer foraging season. Tragically, the remarkable ability of the great baleen whales to store fat has contributed to their undoing, since it has made the harvest for whale oil lucrative, even with the enormous costs of equipping and running floating factories in remote polar seas.

Despite these spectacular blubber stores, the caloric demands of growing baleen whale calves are so enormous, so relentless, that their mothers will lose a quarter to a third of their body weight while nursing. "When I see a large blue whale, when I can see the vertebral processes jutting out of her back, I *know,*" says Diane Gendron, a renowned blue whale expert based at the Interdisciplinary Center of Marine Sciences (CICIMAR), another prestigious marine science institute in La Paz. "I tell my students there must be a calf somewhere. And I'm usually right." Diane is in awe of blue whale mothers. "They raise their calf in six months, and two years after that they have another one, and the calves are *huge!*"

Diane knows the blue whales of the Gulf of California personally, better than anyone else does. "I recognize individuals in the field. That's why I can't stop," she confesses, reflecting back on her twenty-year love affair with the whales. "Every January I can't wait to get out and see who has come back." We are sitting together in her tiny office, our eyes playing peekaboo across the piles of research papers and thick reference books that separate her side of the desk from mine. She reaches forward, pulls out a binder, and begins leafing through the blue whale photo-identification catalog she and her colleagues have created. Each whale is categorized by the color and shape of its dorsal fin (chopped, curved, or triangular). The pigmentation on a blue whale's back is as unique to each animal as your fingerprints are to you. Diane knows thirty or so individuals so well that she immediately recognizes them in the wild. Her students often have to remind her to snap a recent photo of them for the catalog for the benefit of those who don't share her gift.

I ask her to tell me about one of her favorites, and the words spill out quickly, like the boasts of a proud grandmother. It's the story of a female, number 250. Diane was with the noted Mexican photographer and conservationist Patricio Robles Gil the first time she saw 250 with a calf. Patricio was so taken with the little one that he kept asking Diane, "Dónde está la niña?" (Where is the little girl?) So they called the calf Niña. (They even agreed to keep the name when "she" turned out to be a "he"!) Two years later, the same female had a second calf. It had little holes in its skin from either a bacterium or a cookie-cutter shark. So Diane called her "Pinta" (Spot). The third time they saw 250 with a new calf, Diane decided to call it "Santa Maria." She laughs. "They are the names of the three ships Queen Isabella gave Christopher Columbus. So now 250 is called Isabella. And we are hoping she will have a little boy named Columbus next year!"

Recognizing individuals is deeply gratifying to her, but it isn't Diane's primary motivation. Because she knows the exact age of so many blue whales, she can plot their ages of sexual maturity and the frequency of their reproduction. She turns to her computer and pulls up an Excel spreadsheet. "This is the abstract of twenty years of data," she explains. "There are whales that we have known since 1985, and they were already mature then." She points to a line marked 131 and shows me the record for this female's calves—a little one born every two years. A tender smile crosses her face. "Many of these calves are coming back. Some of them are twelve or fifteen years old now. Soon they will likely reach the age when they can give birth, so we'll be able to find out the mean age at which females start to have calves. This is something we don't know now. We are getting an incredible database."

Knowing the specifics of individual whale lives also enables Diane to compile growth rates and other indicators of life history for these enormous animals. In 2005 she and some of her students began taking laser distance measurements and photo identifications of known individuals. They also took biopsy samples. "I feel like I am a doctor and these are my patients," she says. "With the biopsy we know their sex, their length, and if they are pregnant. We know the haplotype. We are doing fatty acid studies—structural and more feeding based. We are doing isotope to see if where they feed in summer is the same place as all of them or not. And we are studying hormones. All of that with one little biopsy. I believe at the end of this year, we'll be able to pluck all this information for each individual to see if they separate into two general groups or just one general group."

The question of whether or not the blue whales that feed in California each summer are members of one population or of two or more distinct populations that travel to different breeding and birthing grounds in the winter is critical. "You need these parameters to be able to say whether or not a population is growing," Diane explains. Blue whales shift their distributions all the time, she tells me, making estimates of population based solely on counts less reliable. If a population is falsely deemed to be growing fast, Diane warns, it could lose its protected status.

There are many unanswered questions in cetacean research. So little is known about many whale and dolphin species and populations. One of the really big mysteries is where blue whales give birth. You'd think it would be easy to spot mothers the length of Boeing 737 airplanes delivering babies as heavy as full-grown hippos. But the ocean holds her secrets close.

Blue whale mothers don't gather in winter calving lagoons the way gray whales do. Nor do they return to traditional feeding grounds each summer. Blue whale mothers depend on a constantly moving food resource, on thick patches of krill (tiny shrimplike creatures). Yet discerning where blue whales most often feed and give birth is vital for blue whale conservation. Blue whales are among the most endangered whales. Scientists estimate that the whaling industry destroyed 97 percent of the world's population. Forty years after the end of commercial whaling, their numbers still haven't rebounded.[3] The largest remnant group in the world, 3,000 individuals, forage off the West Coast between Mexico and British Columbia in the summer and fall.[4] Yet no one knows exactly where this population breeds and how these whales migrate there.

Diane believes that some blue whale calves are probably born in the Gulf of California. "We never see calves born here," she tells me, "but we see females with small calves, so it is quite obvious that if they aren't born here, they are born somewhere close by." Sandy Lanham, who has been flying blue whale surveys with Diane for several years, agrees. She tells me that recently Diane positively identified a female blue whale in the gulf and then respotted it a month later, this time with a very young calf. "I see fin whales that are pretty assuredly giving birth in the northern gulf, and I think blue whales are giving birth there too," Sandy says. "But I am not a scientist. I have no data, nor do I have to." Yet there's little question that she's probably seen more whale births from her aerial vantage point than anyone else. One thing that's nice about not being a scientist, she tells me, is that you can have opinions.

Bruce Mate, director of the Oregon State University Marine Mammal Program, isn't so sure. "We've tracked blue whales from California into the Gulf of California. My own feeling is that it isn't the calving area. The blue whale calves we see there are at least a month and a half old. I've never seen any newborns. So I think they come in there basically to be in a more quiet and restful place."

If finding the blue whale calving region were a game, Diane and Bruce would clearly be playing for the same team. They both want blue whales to be conserved and recognize the need for scientific evidence that will, in turn, help to generate the political will necessary to get these endangered animals, and the habitats and food resources they depend on, protected. They've both devoted their lives to whale research and conservation. But they have different ideas about the best ways to win the game and the levels of risk that should be taken.

While Diane begins with individuals and builds the story of blue whale lives from that perspective, Bruce focuses on entire populations. "The most fundamental pieces of information we need about whale populations is where they live, how many there are, and what's important about those places," he says. "And these fundamentals seem way too basic for most people to think them really important. But our understanding of whale activities is at least a century behind our understanding of land animals." A *century* behind. "Land animals are visible," Bruce explains. "You can see them with binoculars or put radio-tags on them and track them around. If they move, you can move with them. That's not true for the marine animals. They spend most of their lives underwater, and when they're underwater, they're not trackable visually or with conventional radio gear. So it's just that much harder to do these things. And that technical difficulty has kept us at arm's length."

Bruce is the master of bridging that technological gap. His lab pioneered large cetacean tagging in the early 1980s and has been the major innovator in its development and use ever since. Today he directs large-scale, comprehensive, and collaborative marine mammal research programs in most of the world's oceans. As of 2006 he and his colleagues had tagged an astonishing number of blue whales—144. The vast majority, 139 of them, were tagged off California, Oregon, and Baja, and a few off Chile—the "Pacific system," as Bruce describes it.

Each tag is a minicomputer attached to several sensors that measure variables such as pressure, temperature, light level, and salinity. Some tags store the data until a researcher collects the tags and downloads the information in the lab. Other tags transmit their data directly to an ARGOS satellite orbiting overhead when the animal surfaces, or immediately upon detaching from an animal and popping to the surface.

Thanks to the information derived from these tags, Bruce discovered that blue whales range all the way up to Vancouver Island in the summer. And he thinks he knows where they go to give birth. "In the winter we're tracking them all the way down to 8 degrees north," he says. "So not quite to the equator, but way far offshore to a place called the Costa Rica Dome. We think that's one of the key calving areas." Bruce believes the remote nature of the area prevented its discovery until recently and thus fortuitously kept whalers from eliminating these animals as a species. "It wasn't as easy a target as, say, the Pacific lagoons of Baja where gray whales go," he explains.

In many ways, Diane and Bruce's research approaches are complementary. Diane knows, respects, and cares deeply for individual animals and thus helps us feel compelled to conserve them. She recognizes individuals and is able to record the events of their lives over decades. At the same time, conservation decisions are based on what is best for populations, not necessarily individuals. Bruce's work helps us to see blue whales at the scale of a population. Yet a deeper philosophical question lurks beneath conservation research: Under what circumstances should one sacrifice the individual for the good of the group?

Perhaps the term *sacrifice* is too strong, since blue whale research programs must adhere to animal welfare regulations and are not intended to harm individual living whales. Yet any time a researcher follows a whale with a boat or airplane, extracts a blubber sample, or attaches a satellite transmitter, there's an impact on that individual whale, and by extension, on its population, to be considered. Like so many biologists who study threatened or endangered species, Diane is in a painful situation. She knows the wonder and vulnerability of these individuals, these mothers, these endangered species, yet she must decide how intrusive to be in order to protect their population as a whole. Her friend and colleague Sandy shares, "Diane went through agony about deciding whether to dart them; to take a tiny biopsy for DNA analysis. She went through years of not wanting to do that. And she finally decided that she would go ahead and do it. I think she made the right decision."

Diane worries about the impact of whale watchers, photographers, and more insidious threats such as coastal development and underwater noise pollution on blue whales. But she also worries about the impact of her own work and that of other researchers: "I am so concerned about what we do when we study them. I know we disturb them," she tells me. And she has qualms about how satellite tags, for instance, affect the individuals that have been tagged. "The way the tags are working now, they're nailed in, they're designed so they go in, and they don't go out," she says. "We have found whales with bumps with the tag in it. I believe we should start with a few tags and try to see the development of them. And then if it is not really harming, we put some more on."

Diane appreciates the value of satellite tagging programs but is frustrated by what she perceives of as a refusal to take advantage of her knowledge about the targeted animals. "It is crazy, because if you can put a tag on a blue whale, you are obviously in a good position to take a picture to identify it," she says. "We are studying the same animals. I told them, 'Maybe I could tell you if the whales you are tagging are juveniles or whether it is a male or a female so you have a better understanding of who you are looking at.' But they are not interested. They just want to know where blue whales go."

Indeed, Bruce is eager to get as many tags on whales as quickly as possible. "Last year we had tags on five populations of different species in four different oceans," he says. "If I could click my heels together three times and make all the tags I want appear, and salaries, and travel, and ship time and all that, we'd be doing four times as much as we do now." From a population perspective, each whale is piece of a jigsaw puzzle that collectively reveals the picture of what a population does. To Diane, each whale is unique. "What if that whale is a female?" she asks. "Let's say it is a female and we know it is because we have seen her with a calf. The tag might prevent her from producing the number of calves she was supposed to. Or maybe not. But at least consider the possibility." She pauses and looks me straight in the eye. "It's a woman thing, I guess."

Cetacean research is a young field. Like many scientific fields, men dominate it. "In whale research, ten years ago," Diane tells me, "you would see all those famous researchers with all those young beautiful girls. You still see that, but I think it is shifting. In a couple of years, maybe five, there will be more incredibly good women researchers." I ask Diane whether she thinks the kinds of questions she asks as a scientist, the approach she takes, are influenced by the fact that she is a woman. "I think so," she answers. "This is very personal, but this feeling of knowing the animals to protect them better, I think all this concern comes naturally to me because I'm a woman."

Is it really an issue of gender, I wonder? Or is empathy with individual animals a personal quality? When I ask Diane who else I should speak with to really understand blue whales, the name of a boat captain springs from her lips. "He knows them so well. I don't know anybody else who can get so close to a blue whale and not harass them." Diane describes this captain's ability to see whales long before anyone else. How he proceeds carefully, slowly, until a blue whale mother and calf is right there and the boat is in neutral. "He is my best friend. He is so sensitive. Sometimes too much," she says. "He has been doing whale watching since 1982. When he sees a big boat or a panga chasing a blue whale he gets really angry."

Teasing out the influence of gender on any issue is complicated, and in the case of conservation biology, political dimensions of science and conservation policy factor into the equation. Diane works for a Mexican university. Bruce is based in the United States. The whales move through the waters of both nations. What influence does U.S. or Mexican policy on marine mammal protection have on the kinds of evidence researchers must collect and thus on the types of research programs they must carry out to generate the political will to conserve?

The Gulf of California is a Mexican whale refuge. According to Jorge Urban, "The law in Mexico defines sanctuaries as places you cannot whale. It is a refuge from whaling." The designation has no direct impact on major whale conservation issues such as underwater noise or entanglement in fishing nets. The situation is much the same in the United States. "The U.S. policy is we protect marine mammals," Bruce tells me. "The reality is, our policy is we don't *kill* marine mammals intentionally."

When it comes to conserving endangered or threatened whales and dolphins, it's not enough. "If you're going to protect something you have to know something about them," says Bruce. "For many species we don't even know who, where, what, and when, and that's really important because you can't assess the population unless you know where and when they occur. You can't protect the habitat unless you know what it is. And you certainly can't identify threats if you don't know what the geographic region is."

The same is true in Mexico. Mercedes Eugenia Guerrero Ruiz is currently working with Jorge on a program to document all the species of marine mammals that inhabit Mexican waters for CONABIO—the Mexican National

Commission for the Knowledge and Use of Biodiversity. "In general terms, we can say that what we know about cetaceans in Mexico, it's nothing" she laments. "Certain species have more political importance. Gray whales, for example, have given recognition to our country that we care, that we protect them. But most of the species that we have here, we have no idea. You can speak in general terms about these species, but probably the population of false killer whales that inhabits the Gulf of California behaves completely different from the ones from somewhere else."

There's the rub. A whale is not a whale is not a whale. If there is one takeaway lesson from the combined research of whale biologists in Mexico, the United States, and around the world, it is that whale populations living in different places lead different lives. Even in the same place, different whale species engage in different cultural norms. The more researchers look, the more they find that individual whales are unique. Conservation policies demand evidence that looks across individual variation to find what is common to most. That's why Bruce is so driven to expand the satellite-tagging program. He is looking for commonality. He is looking for the holy grail of conservation, the hot spots in the ocean that are most vital to a broad range of individuals, populations, and even species.

While you are reading this, a blue whale is swimming in the North Pacific. Each time it surfaces, information from the satellite tag on its back is beamed up to the heavens and bounced back to earth by a satellite. The same is true for hundreds of bluefin tuna, sea turtles, white shark, billfish, and other whale species. "You look out on the water, and most people just see a flat surface," Bruce explains. "They have no idea whether it is desert-type productivity or grasslands or forest. They can't monitor it for themselves with their eyes. So these kinds of assessments of characterizing the habitat, of overlap of multiple species use, is important." Under the umbrella of the Tagging of Pacific Pelagics (TOPP)[5] joint research program, Bruce is working with more than a dozen scientists to chart the movements of twenty-one species in the Eastern Pacific to determine how similar or different their needs are.

Meanwhile, Diane plans to spend her next sabbatical year in the company of her beloved blue whales. Following in the intimate tradition of Jane Goodall's work with chimpanzees and Cynthia Moss's with elephants, she will follow individual whales, one whale each day, by herself. "I don't need to get close. I just need to know who it is and to follow it and to see what happens. Then I am sure we will understand things we never thought possible."

3

Looking for Whales in All the Wrong Places

These past few years I've been thrust, like so many other parents, into re-discovering nature in my own neighborhood. Not necessarily because I wanted to, but because the two-year-old I was walking would throw herself onto the ground and simply not budge until the ant crawling ever so slowly made it *all* the way across the sidewalk. Why such "moments of wonder," as an old naturalist friend of mine calls them, typically occur when I am desperate to get another child to school or myself to a meeting without any stains on my clothing is unknown. Perhaps it's because we always seem to be rushing to beat some bell.

But in quieter, less hurried moments, I really envy young children, who, given the time and the freedom to explore, seem better able than most of us to see things for themselves. My son, Kip, finds life everywhere—slender salamanders in the gravel patches of our driveway, crabs in the litter-strewn rocks bordering a boat ramp. He finds them because he doesn't know enough to know that these aren't the best places to locate animals. He takes the time to look. When we snorkel together, he often doesn't like others to tell him what to watch for, preferring to float in the water, staring intently at tiny shrimp larvae until the convulsions of his shivering body become so great he can no longer focus and is forced ashore to heat up just enough to go back out.

Bruce Mate finds a lot of whales in places that aren't good for finding whales. It's not that he doesn't know the best places to look; it's that he's acutely aware that finding whales is a kind of self-fulfilling prophecy. You know where whales go, so that's where you look for them. You find them there, and that reinforces your belief that that's where whales go.

The trouble is, this technique tells you nothing about the places you haven't looked, or what whales do in the places where you don't see them. And that's where the great surprises lie. "There isn't a single population of whales we've worked with," Bruce explains, "that when we went in and tagged the animals for the very first time, we didn't learn things that were dramatically different from our preconceived notions. Like right whales *(Eubalaena glacialis)* in the North Atlantic. We thought they were slow-moving, near-shore, surface-skimming feeders that spent the summer in the Bay of Fundy. Instead, we found that they move fast, cover long distances, can dive deep, and routinely leave the Bay of Fundy. The reality was 180 degrees off our original notion."

What's humbling about Bruce's example is that it's easy to understand how such misconceptions occur. To study a population of whales, you need to be able to identify individuals. And that's no easy proposition. In the case of right whales, the discovery that each whale has a distinct pattern of markings on the top of its head spawned the long and arduous task of photographing individual right whales by aerial survey. The photos are compiled into a photo catalog. Each individual has its own page with a collection of photographs taken over the years and its own name or number. Much like in a family photo album, there are many photos of some individuals and barely any of others. Researchers use the resulting mug shots to identify whales in the field and thus to build the life stories of individuals: who they spend time with, where they are seen, whether they have a calf with them, and so on.

Today photo IDs are the number-one technique used by researchers to answer fundamental questions about where the world's more than eighty species of whales, dolphins, and porpoises live; how many there are; and the structure of their pods, or groups. You've probably seen pictures of whale researchers standing on the bows of boats shooting photographs of whales, and more specifically, of whale tails or dorsal (back) fins, using high-magnification camera lenses. They can tell individual humpback whales *(Megaptera novae-*

angliae) apart by looking at the color patterns and overall shapes of their tail flukes. Back color distinguishes blue whales. With killer whales, it's the size and shape of their dorsal fins and the shape of the distinctive saddle patch behind the fin that distinguishes who's who.

It was photo IDs of right whales taken in the Bay of Fundy over the course of many summers that led to the idea that right whales spend their summers and early falls in that area. But the reality, according to the information that Bruce gathered by satellite-tagging individual right whales, is quite different: "Right whales frequently come back to the Bay of Fundy over the course of the summer," he explains, "but they make these big looping trips out to the Scotian Shelf, east of Nova Scotia, or into the Gulf of Maine or sometimes farther." Researchers only took photo-ID shots of the whales when they were in the bay, rendering invisible, and therefore nonexistent, the hundreds of miles the whales that traveled between photos.

Don Croll and Bernie Tershy have been studying the ecology of marine mammals, often in the Gulf of California, and seabirds for decades. They lead the Croll-Tershy conservation biology and marine ecology lab at the University of California, Santa Cruz, and are cofounders of Island Conservation, a science-driven nonprofit organization dedicated to preventing extinctions and to protecting natural ecological and evolutionary processes on islands. Bernie tells a funny story about a graduate student friend of his who had a professor with a horse ranch. The horses were completely hidden behind a fence, and every once in a while, one of them would poke the tip of its nose above the fence line. You could tell individuals apart by the color of their noses. His friend took a lot of pictures of horses doing that. And then he concocted a theory about what the horses were doing when you couldn't see them. "It's a good analogy for whale research," he explains. "Basically, with whales, we spend a lot of time watching animals breathe and not looking at what they are doing the rest of the time."

Don believes that whale research has been terribly constrained by its methods. "I am amazed at how long people have just been photo ID-ing whales," he says. "We're still doing a lot of the same things that we were doing twenty-five years ago. How much more are we learning?" Bruce concurs. He's worked with Don and Bernie on blue whale research and is well aware of the limitations of depending exclusively on photo identifications. He emphasizes this point by referencing humpbacks in Hawaii, a population that has been extensively studied by researchers and is admired by thousands of eager whale watchers each winter. "The first time we satellite-tracked a humpback whale from Hawaii to its summer feeding ground," he tells me, "everybody at that time thought Hawaiian humpbacks all went to Southeast Alaska. That's because that's where all the fluke photographs were being taken during the summer feeding season. And the very first one went to the Kamchatka Peninsula of Russia. We didn't even know that whales from Russia came to Hawaii. And why is that? There's nobody taking whale fluke photographs in Russia."

Part of the problem, according to Don, is that studies of whales are geographically determined. Researchers study places where they have access, places where they can launch boats and mount their studies. Or where their university is located. In the Gulf of California, for example, Diane Gendron and Jorge Urban both work out of research centers in La Paz. "And so Diane and Jorge study whales in the Bay of La Paz. Maybe up to Loreto," Don says. "So now there are a lot of marine mammal people in La Paz. Thus, a lot of what we understand is from the study of animals down in the southern gulf. There is a lot of gulf, but not that many places where people study whales."

Bernie contrasts La Paz, a major tourism center, to Mexico's isolated southern coastal states of Sonora and Sinaloa. "There's a whole bunch of mangrove swamps, tons of insects, a lot of drug running, kind of tough places to be and no latte shops or anything," says Bernie, "and there's not much research going on there." Both these regions of the Gulf of California are really important areas for whales, but only one is well studied. "It's the same for the Bering Sea and some of these awful places that either because of cost or safety or politics or boundary issues have been off-limits," Bruce explains.

It's not just location that keeps researchers from finding whales. There's also the issue of equipment. Most whale researchers working in the Gulf of

California study whales from pangas, the smallest and least expensive of the fishing boats commonly found in Mexican waters. The money just isn't there for bigger ships. But when the whales being studied are double or triple the size of the boat, it limits perspective. Bernie sounds like a car buff tracing his ascent from beater to BMW as he describes his research history through boats: "I started studying whales out of a thirteen-foot boat. It had a 20-horsepower engine. And then when I actually got a panga, a twenty-foot boat with a 55-horsepower engine, it really expanded my range. All of a sudden I started learning all these new things about whales."

It is, he tells me, like trying to understand my life by studying me in my driveway. Sure, I spend some time there, but I also go to work and care for my kids and travel to Mexico and write books and . . . you get the idea. "We were trying to figure out what these whales do at a spatial scale that is completely inappropriate for them," he says. "Every time I expanded my geographic perspective, I realized there's still a lot to learn about what was driving them." These days, Bernie and Don look at a blue whale population pretty much across its entire range, at least for some of the individuals. "The scale that we're looking at is appropriate to the scale in which they move, but they're not spatial scales that we're comfortable with as humans," Don tells me. "Once we kind of got our heads around the fact that the backyard of a blue whale extends from the Costa Rica Dome up to the Gulf of Alaska, we started to get a better handle on what these animals are doing."

Can a human being ever comprehend whale scale, I wonder? What must the world feel like to an animal that can travel a hundred miles as easily as we walk a city block? I think of Sandy Lanham, the pilot who flies aerial surveys for researchers, and imagine that she, if anyone, might look down and see the Gulf of California in whale-size proportions. Yet even her perspective is limited by human constraints.

When whale researchers talk about aerial surveys, Sandy tells me, they aren't talking about crossing the gulf once. They're talking about sixty hours of flying out across the gulf in a risky series of flights. She and Diane began flying surveys together in the early 1990s. "The original work that Diane and

I did, we stayed within twenty-five or thirty miles of the coast. We surveyed areas that Diane could return to in a panga. It was cheaper," Sandy says. "But it was definitely a safety issue too. They say the number-one killer of marine mammal researchers is aerial surveys. It's circling over whales and getting too involved in what the whale is doing and losing track of your air speed."

Doc White, marine mammal photographer, knows the treacherous side of getting too focused on a whale. He photographs whales underwater and on one memorable occasion fell prey to the allure of incredibly clear water and a swimming humpback whale. "I looked down, and there he was hanging, inverted, tail up," says Doc. "An absolutely gorgeous situation, and I was diving with a snorkel. I dove on the animal and started taking images. I took about twelve shots and realized I needed to get to the surface. I looked up, and realized I was down about seventy feet. I started swimming for the surface, my safety dive partner was at the surface, and I took my weight belt off and held it in my hand so that if I did have a shallow water blackout, I would at least drop the belt, and hopefully my wetsuit would give me enough buoyancy to get to the surface. It was not smart. You get too involved, and all of a sudden you get in trouble."

Despite the risks, Diane wasn't content to stick to the shores. She felt driven to explore the Gulf of California in blue whale proportions. Though she still isn't sure why, Sandy finally agreed to grant Diane's wish. "We started doing these massive long surveys where we would fly from one side of the gulf to the other." The results were magical and immediate. "Diane has discovered that there is an enormous population, the biggest population of blue whales we know of, out in the very middle of the northern gulf," says Sandy. "We've seen them the last two years in a row. Probably we didn't find them before because I wouldn't fly across the gulf."

Doc spends a lot of time around whales. But he wouldn't put it that way. "When you say you're around these animals for long periods of time, it's actually a misnomer," he corrects me. "You're actually *looking* for them for long periods of time. Scheduling a date with cetaceans doesn't work. You've got to have time, and you're going to be skunked. Watching whales is like fish-

ing. They don't call it catching; they call it fishing." There are, of course, surprises, times when whales seek you out. "We were in San Ignacio Lagoon,"
Doc says, "very shallow, anchored at night. After the evening meal we heard
these blows, and there were two gray whales on the surface on the swim step,
using the lee that the boat created, and the tidal current, and just laying there.
And in the middle of the night, on numerous occasions I'd wake up thinking we had slipped anchor and run aground, and it was gray whales under
the boat rubbing against the hull. Scared me to death!"

I ask Doc to share his best advice about how to see whales. "Time," he tells
me, "is the bottom line. You've got to be out there. You won't see them in bad
weather. You can't see them at night. Most of the time they're hidden beneath
the surface. It really cuts down on how much we know about them."

It's a fascinating concept. What we know of whales we extrapolate from
the tiny glimpses researchers get at the water surface, and even then, only
by looking in the same old places. It makes me think of the line in that old
Waylon Jennings hit, "looking for love in all the wrong places." The whales
are out there, somewhere, but like the way little kids play hide-and-seek, we
can't find them because we only look in the places where we've found them
before. Places where we are comfortable. Places where we aren't afraid to look.

4

Resident Aliens?

It's a conversation I know by heart. I bet you know a version of it too. This time it's being told by a farmer's wife from outside Boise, Idaho; she leans over to chat while we're waiting for our plane to be repaired at the airport in Guaymas, Mexico. "We're just here to check out our friend's new condo development," she tells me. "I'm in real estate too, and I made more money in the last six years than my husband did his whole life farming, and I only work three months a year!" She adjusts her heavy frame in the tiny molded plastic seat and examines a bunion on her sandaled foot. "It's a great spot. You can see the ocean from every room." She passes me a crumpled brochure with a picture of the condo and presses a manicured finger on a panoramic shot of a villa nestled against the Gulf of California, or as Steinbeck enthusiasts refer to it, the Sea of Cortez. "That's where they shot the *Mask of Zorro.*" I scan the text and read: "San Carlos is only a 250-mile, four-hour drive from the Arizona border on Highway 15, a four-lane divided highway." The developers know their market. Seventy percent of the residents of this beautiful resort community in Sonora come from the United States and Canada.

Conversations about real estate investment and the "best" places to live are the talk of our time. So many of my neighbors, even those young enough to have small children, made a killing in the California real estate market, and they're now actively pursuing the dream of owning beachfront property someplace where the cost of living is lower, health care is cheaper, the weather is warmer, and the pace of life is less frantic. Mexico's Puerto Peñasco, in the state of Sonora, and Loreto Bay, in Baja California Sur, are two of the hottest spots. Situated, respectively, on the east and west shores of the Gulf of California, these sleepy fishing villages aren't much farther away than many people's summer cottages and have the added advantage of gorgeous swimming beaches and a family-friendly culture. But they won't remain sleepy for long. Changes to Mexican laws in the wake of the 1994 North American Free Trade Agreement (NAFTA) make it easier than ever for Canadians and Americans to reside on Mexican beachfront property. According to the international market development division of the National Association of Realtors, between 500,000 and 1.5 million Americans now have homes in Mexico.[1] "There is no real estate hotter than Baja," says Jim Grogan, developer and CEO of Loreto Bay, a seaside resort development near Loreto. "Baby boomers are seeking a new frontier."

The idea that people can choose to live anywhere, that they can just pick up and move to a new frontier, is so commonplace it's easy to forget what a uniquely human trait this is. It seems logical that other species, particularly those like large baleen whales, must also travel the world in search of new sources of food or mates. But scientific evidence now suggests that some populations of whale species stick much closer to home than anyone once thought.

Fin whales *(Balaenoptera physalus),* for instance, are the second largest animals in the world. A large male can reach a length of eighty-five feet. Even at birth, a fin whale weighs more than 6,600 pounds—more than half the weight of a fully grown Asian elephant. And they're fast. Nicknamed the "greyhounds of the sea," these giants travel at speeds of twenty miles an hour, covering as much as 180 miles in a single day. Fin whales, like most large baleen whales, are migratory. They travel poleward each summer to gorge in the seasonally available, nutrient-rich waters.

Yet in the turquoise waters of the Gulf of California, scientists have made a most unusual discovery: a population of 400 fin whales that don't migrate. "A fin whale can swim in and out of the whole length of the Gulf of California in six days, or something like that," says Bernie Tershy. "What's exciting about the individuals in this population is that they don't. They reside year-round in the Gulf of California."

They are the only known population of fin whales in the world to do this. (Another claim of a resident population of fin whales living in the Mediterranean is currently under investigation.) According to Juan Pablo Gallo Reynoso of the Research Center in Feeding and Development, in Guaymas, some individuals in the gulf have been residents for decades: "We have seven individuals in the population that we have photos of from twenty-three years ago." Genetic evidence analyzed by a team of U.S., Mexican, and British researchers recently confirmed that the Gulf of California fin whales are evolutionarily distinct and represent a highly isolated population that warrants special conservation measures, given its small size and unique composition.[2]

So how do these enormous whales survive without migrating, when others of their species have to travel to find enough food? The answer may have more to do with the Gulf of California than with the whales. The gulf is one of the richest marine ecosystems in the world. Almost all the major oceanographic processes occurring in the planet's oceans exist in this one special sea, guaranteeing accessibility to nutrient-rich water year-round. In 2005 UNESCO awarded World Heritage Site status to the gulf's islands and protected areas in recognition of their extraordinary biological diversity.

"Fin whales stay in the gulf because they can exploit different levels of food," says Diane Gendron of CICIMAR in La Paz. In winter they feast on krill, just as their larger cousins, the blue whales, do. In the summer, however, the biomass is very low in the gulf, and that's why blue whales leave. It's much more profitable to feed off the California coast in that season. Yet the fin whales stay. Figuring out what they're eating isn't easy. Traditionally, scientists determined whale diets by sorting through the smelly gut contents of dead whales. Today Diane and her students collect sloughed-off skin from living whales or extract small skin samples, using a technique called stable isotope analysis, to examine the storage capacity of the whale's skin for chemical signatures. The chemical signature of a whale that eats fish is different from that of one who eats invertebrates. They found that fin whales shift their feeding to a higher trophic level, or feeding position, in the food chain during the summer. In other words, they switch to fish.

This fascinating case of nonmigratory whales illustrates one of the most significant trends in whale research—the discovery that not all whales, even those of the same species, are the same. The fin whales of the Gulf of California also serve to remind us that not all places in the ocean are equal. It takes, as Bernie puts it, "a happy set of circumstances" of geology and oceanography and latitude and temperature to create the rich year-round productivity of the Gulf of California. Most of the ocean is, in fact, a "desert," in the sense of having very low densities of life. Within that vast expanse, particular places are rich with certain kinds of food during certain seasons. Only a few truly remarkable places, like the Gulf of California, are blessed with the conditions to support such an extraordinary variety and abundance of undersea life throughout the year.

The Gulf of California is the best place in the world to see the greatest diversity of whale species. And it turns out that it's not just because of the food. There's another reason why whales come, or in the case of fin whales, stay. It's quiet. Fin whales, blue whales, humpback whales, Bryde's whales *(Balaenoptera edeni),* and sperm whales all come to the gulf to breed and give birth to their young. Diane believes it's the relative peacefulness of the waters that makes the Gulf of California a globally significant whale nursery. Drawing on the example of the gray whale breeding and calving lagoons on Baja's Pacific Coast, she says, "If Baja were to be as developed as 20 percent of California, gray whales would not be as healthy as they are now. Why? Because the lagoons where they come are still the same as they were two thousand years ago. San Diego was once a nursing area for gray whales. It is easy to understand why they don't go back there anymore."

Nursing mothers and newborns need peace and quiet. But when it comes to whales, every member of a population, regardless of gender or age, depends on it. A few years back Christopher Clark, a leading bioacoustics expert at Cornell University, traveled to Loreto Bay, where fin whales aggregate to gorge on swarms of krill. He and his colleagues intended to test new acoustic instruments that they hoped would enable them to locate, track, and monitor individual fin whales. The test was a success, but what they discovered was even more compelling: some fin whales spend a lot of time singing during the krill feast.

Chris and his team revealed the identity of the singers by using special computer software installed onboard a converted fishing boat that was towing an array of hydrophones. Once they located a singer, the researchers collected a small skin sample from it. Genetic analysis of the skin biopsy samples revealed the sex of the whales. All the singers were male. Chris believes the males sing courtship songs using frequencies far below the range of human hearing.

"If there is one place I want to go back and do some more fieldwork in, it's right there in Loreto and all those beautiful islands," says Chris. "There'd be a nice big high-density patch of krill, and male fin whales would set up and sing next to these patches of food. An animal would come to the surface, and the song would stop and he'd breathe, and then suddenly he'd dive, and about ten seconds later the song would start again. Later on, processing the data, we found cases where a male singer would 'leave its perch' and come out singing toward an intruding male. It had all the signatures of a defensive, local territorial behavior—like, OK, this is my territory, I'm singing here, go away. I don't mind sharing some of this, but I've actually got a few females in here."

According to Chris, the very low-frequency courtship songs of fin whales, like those of blue whales, can travel hundreds, if not thousands, of miles underwater. But so can very low-frequency, human-made noises, which have increased dramatically in the last hundred years of motorized shipping. Chris is concerned that the artificial racket created by ships and other human sources could be interfering with whale reproduction and population recovery. "Twenty to twenty-five million years of evolution are being undone in a hundred years," Chris says. "There are hundred-year-old whales alive today who can probably remember when the ocean was a much quieter place and they could communicate with colleagues across grand expanses of ocean." If, as a growing body of research suggests, the great whales rely on long-distance songs to find each other for mating, the recovery rate of whale populations from past exploitation could be impeded by low-frequency sounds generated by human activities such as commercial shipping, military sonar, seismic surveys—and coastal development.

"I am trying to use the blue whale and the research we do to conserve the Gulf of California," Diane says. "If we don't do it, we will lose one of the last pristine areas in the world. And why is it pristine? Because Baja is a desert. There is no water. Development was not possible until now." The connections between luxury condos and the survivorship of a baby blue whale aren't obvious, but Diane sees them clearly. "Development close to shore means more boats, more noise, more everything. More people, more water, more power. How are you going to make the electricity to pump the salt out of the saltwater in the desalination plants? With petrol or natural gas. Baja is almost an island. It is so far from California. How do you bring petrol down from the states or across from Mexicali? It comes by boat. It means big ships and noise and activity in the water. That's the biggest issue for the whales."

"If you could live anywhere, where would you choose?" So reads the first line of the real estate brochure for the Villages of Loreto Bay, the largest real estate deal ever made by the Mexican government and Baja's most ambitious housing project to date. Situated on eight thousand acres along the shores of the Gulf of California, the $3 billion project will be built in phases over twelve to fifteen years. Plans call for a mixed-used development featuring six thousand homes, boutique resort hotels, two eighteen-hole championship golf courses, a luxurious beach club and spa, a tennis center, a marina, and sportfishing center, along with restaurants, boutiques, galleries, and a produce market. Five thousand acres will be maintained as a natural preserve for hiking, cycling, and organic farming. Loreto's population is expected to grow in this time from 15,000 to 120,000.[3]

Diane's concerns about the impact of this development on the whales and the marine ecosystems of Loreto Bay are echoed by a group of professors at Harvard University's design school, which released a 2005 report called "Loreto Alternative Futures."[4] The report confirms that desalination is the only option to support the growth and warns of the ecological, visual, social, and economic risks posed by the Loreto Bay plan.

"If you look at the economy in the Gulf of California," says Exequiel Ezcurra, provost of the San Diego Natural History Museum and director of its Biodiversity Research Center of the Californias (BRCC), "it is the tourism sector that offers any type of expectation for growth. One of the big discussions for Baja, and the Gulf of California in general, is not whether there will be tourism, but what kinds of activities should be promoted." Exequiel contrasts the ecologically sensitive approach to tourism along the Pacific Coast of Baja in the gray whale breeding and calving lagoons to the recent megadevelopment of Los Cabos. "It's an environmental nightmare of gargantuan proportions," he says. "It is the big hotels that demand huge amounts of energy, big tanks of oil moving almost every day from mainland Mexico to Los Cabos to sustain the huge amount of air conditioning and energy-intensive installations. They demand huge amounts of water for their golf courses. Golf courses in the desert are one of the stupidest ways of using water you can imagine. And in general it is a resource-intensive activity that, believe it or not, needs to be subsidized by the federal government so they make ends meet."

The nonmigrating fin whales of the Gulf of California present an intriguing juxtaposition: the impacts of people who migrate on whales that do not. Dr. Ezcurra is a highly regarded scientist who is well aware of the "enemies of progress" image so often attributed to those who raise environmental concerns. In 2001 President Vicente Fox appointed him president of the National Institute of Ecology (INE) for Mexico. He currently leads an initiative focused on understanding and managing the Gulf of California as a whole, single ecosystem, including the people who live and work there.

"How," I ask him, "does one deal with the conundrum of people eager and able to move someplace *else* and their impact on whales that are entirely dependent on someplace *special?*" He takes time to consider his reply: "In many ways the Gulf of California is like a microcosm of the big questions that human beings will truly have to deal with in the next decades. What sort of future do we actually want? Do we want a future with big hotels using water unsustainably, consuming huge amounts of fossil fuels, and creating immense emissions of greenhouse gases? Where are we going as a biological species? Decisions about the Gulf of California can be considered some sort of trial of the way we make decisions in the broader, global context."

One thing's for sure. The whales of Mexico's Gulf of California can already hear the conversation.

5

How to Make a Really Rich Sea

Steve Webster is in love with the Gulf of California. "It's one of those magical places," he says. "You know you belong there when you get there. In fact, it happens to me about the time I stop for gas and pick up a lobster burrito in El Rosario." Steve is an accomplished marine biologist and director of education emeritus at the Monterey Bay Aquarium. He felt the pull of the place the first time he drove down a little stretch of rough road south of San Felipe to Bahía de Gonzaga. It was a Christmas trip with a couple of grad students from Stanford in 1968. "It snowed at sea level," says Steve. "It was a very cold winter, and I got hooked right then."

It is an enchanting image. Snow in a Mexican desert, a fitting first impression of a region defined by unexpected contrasts. If the Gulf of California were a little farther south, it would be all tropical. If it was a little bit farther north, it would be all temperate. But it's north-south. And it's convergent between the temperate and subtropics. If you were to steam across the Pacific in a boat, you could go for days without seeing much change in depth or temperature or oceanography. In the Gulf of California, however, you find amazing variety. The surface waters at the mouth, toward the southern tip of Baja, remain tropical year-round. Those in the north experience a spectacular seasonal transformation. In the summer they soar to 80 degrees Fahrenheit. These bathwater temperatures unleash a seasonal surge of tropical fish: brilliantly colored angels, butterflies, and damsels. Each winter, many of these fish vanish. "I don't know if they go deeper or if they go south," Steve says. "Maybe a little of each. But they certainly must move around because the numbers are quite different in the two seasons." Clearly, water temperature is the culprit, dipping as it does to the mid-50s each winter. Contrast that amazing temperature change to the water temperature gradient between Alaska and Central California, which is less than 10 degrees. "The temperature difference between the southern gulf—from the mouth of the gulf—and the northern gulf in winter," he laughs, "is just *ridiculous* for a marine area."

This distinctive north-south orientation is a key reason the Gulf of California ranks among the richest, most species diverse places in the world. It's got a big latitudinal and big temperature range. But that's not the only benefit of being north-south. "Anytime you see a north-south facing large coastal basin on this planet," Steve explains, "you know you're going to get extreme tides at the head of that body of water." The Gulf of California dead-ends at the northern end. "It's like a bathtub," he says. "The water goes in and it kind of sloshes up against the other end, and it goes up. You get these huge tides." Indeed, the tides in the northern end of the Gulf of California have the third highest fluctuations on earth, behind the Cook Inlet in Alaska and the Bay of Fundy in Canada. In San Felipe the shoreline drop between high and low tide can be as much as twenty-two vertical feet. And at the mouth of the Colorado River that tidal fluctuation can be over thirty feet. To put this extreme tidal range in real estate terms, a house in the northern gulf that's half a mile inland at low tide is a waterfront property when the tide is high.

For a marine invertebrate enthusiast like Steve, exploring a half-mile intertidal zone (the area between tidemarks) is nirvana. "The tidal excursion is so high, it just expands whatever tidal zone is there at the head of the gulf," he tells me. "Combine that with being in a desert that's extremely hot in the summer and quite chilly in the winter, and it means that the critters are going to have to contend with some really extraordinary seasonal and even daily changes between day and night, between high and low tide, between summer and winter." The result is an extraordinary diversity of marine habitats and therefore, an extraordinary diversity of marine animals.

Six decades earlier, John Steinbeck and his biologist friend Ed (Doc) Ricketts were drawn to the diversity of life that exists between these tides. Steinbeck had just finished writing *The Grapes of Wrath* and needed a break. He didn't want to think about the looming war, nor was he ready to face the consequences of his recent literary success or his failing marriage. "Now I am battered with uncertainties," he wrote in his journal on October 16, 1939. "That part of my life that made *The Grapes* is over. I have one little job to do for the government, and then I can be born again. Must be. I have to go to new sources and find new roots. . . . I don't quite know what conception is. But I know it will be found in the tide pools and on a microscope slide rather than in men."[1]

Steinbeck believed that science held the promise for a world he felt badly in need of revision. In his friend Ed, he found a biologist ahead of his time. Ed was the kind of guy who would decide to walk from Chicago to the Deep South and back again because he wanted to explore the world close-up. A self-trained marine biologist and owner of a biological supply house on the legendary Cannery Row, he looked at life from an ecological viewpoint at a time when concepts such as niche, habitat, food chains, and predator-prey relationships were still uncommon.

The rich waters and desolate landscape of the Gulf of California promised an escape from the troubles of home. The journey was to be a new start, a mission of renewal. When they set off from Monterey Bay for the Gulf of California in 1940, they planned to conduct the most extensive and objective study of invertebrate distribution ever undertaken in Mexico. "Cortez was for John Steinbeck what the Galapagos was for Darwin: a pristine panorama of the natural world, perfect for the illustration of profound interpretations of biology," writes literary scholar Brian Railsback.[2] For forty-one days, they worked their way along four thousand miles of coastline; the absence of even a shipboard radio freed them from news of the impending war. With knees licked raw by the salt of the tide pools, Steinbeck and Ricketts recorded more than 550 species of marine life. All from the seashore.

The pair never explored beyond the coast. "I'm not sure they even mention a whale," Steve says, referring to the *The Log from the Sea of Cortez*, the book that emerged from the expedition.[3] "They mention a few big fish, like the manta ray that one of the crew was trying to harpoon. But very little mention of seabirds, whales, sea lions, or dolphins. It's as though they were so focused on the intertidal that they just kind of ignored everything else." Journeying as they did in the days before sonar and scuba, Steinbeck and Ricketts could not have imagined the remarkably deep seamounts and underwater canyons that exist beneath the surface or the geological processes that cause this undersea world to shift, buckle, and break. They did not know how depth, temperature, availability of light and nutrients, and characteristics of the bedrock and sediments influence seafloor ecosystems, and how these in turn influence and are affected by currents, tides, storms, and other above-water events.

Ironically, World War II would bring with it the technical advances to do so; innovations so substantial, it's impossible to contemplate ocean exploration without them. Like schools of robot fish, undersea gliders and other robotic undersea vehicles, along with other instruments carried on ships, airplanes, satellites, buoys, and drifters, now provide oceanographers with new ways of seeing and hearing the ocean in unprecedented detail. Undersea technology is a testament to the power of science that Steinbeck prophesied.

In the past decade, remotely operated deep-sea vehicles (ROVs) from the Monterey Bay Aquarium Research Institute "boldly go where no man has gone before," penetrating the mile-deep Midriff channels that slice through the bottom of the Gulf of California, probing the relatively shallow regions of the northern gulf that seldom run deeper than six hundred feet, and descending into canyons in the southern gulf that reach depths of ten thousand feet.

If you were lucky enough to ride along in an ROV as it explores the undersea ridges and canyons of the gulf, you'd discover that this magnificent giant of a sea is just a baby. Indeed, it is the world's youngest sea, born 5 million years ago when the San Andreas Fault split the Baja Peninsula from Mexico. The shifting dimensions of its geological background further account for the sea's current diversity. "It's clear that when the isthmus was below sea level you had some of the tropical Atlantic critters moving back and forth and up into that area," Steve says. "You see a number of fish and invertebrate species, like the sergeant major damsel fishes, and soft corals, gorgonians, as well as hydrozoans, that had their roots in the Atlantic Ocean." At the same time, in the tropical regions around Cabo Pulmo on the Eastern Cape, you get Pacific species such as crown of thorns sea stars, yellow-bellied sea snakes, and Moorish idols. And you also get species moving northward off Ecuador and the Galapagos Islands and down from the California region. "There are California sheephead as well as gulf sheephead in there," Steve explains. "You've got a couple of species of sea bass that very likely have their roots in Southern California. Biogeographically, it's in a place where you get elements coming in from a number of different directions over a variety of time periods."

It's this wonderful combination of north-south orientation, extreme temperature differences, deep and shallow seas, geological history, and, of course, monster tides, that create such an enormously rich sea. The Gulf of California is home to 170 species of seabirds, 3,000 species of invertebrates, and 875 species of fish. The nine hundred or more islands and islets in the gulf serve as important nesting sites for migratory and residential birds. It is a critical breeding, feeding, and nursery ground for thirty-two species of marine mammals. The coral and rocky reefs that dot its shallow waters are home to a wide variety of endemic species—species found nowhere else on the planet.

The statistics are superlative, but the best way to hear the full symphony of the Gulf of California is to don a mask and immerse yourself in it. Look at the desert island above water, and then look at the ocean productivity below. It doesn't matter where you are, near shore, in the subtidal zone, or floating above the abyss, it's all about this one great ocean system. Everything you see is intimately connected. "A man looking at reality brings his own limitations to the world," Steinbeck said. "If he has strength and energy of mind the tide pool stretches both ways, digs back to the electrons and leaps space into the universe and fights out of the moment into no-conceptual time. Then ecology has a synonym which is ALL."[4]

6

Popular Mechanics

I love the feel of our rented house in La Paz. The air is so hot, so dry, that we hang wet clothes at midnight and pull them stiff from the line before sunrise. Water is precious and so hard to come by that we shower together, shivering as the drops remove all evidence of our salt-and-sand adventures in the Gulf of California. We memorize the 1950s jingle that signals the arrival of the water truck and race to greet it with the joyful sense of anticipation previously reserved for the ice cream van. There is the stink of gas seeping from the stove, and our feet stick to the cracked linoleum, onto which juice from *mandarinas,* fresh from the tree in the courtyard, spilled over the bent edge of the fruit press. We lie on chlorine-speckled sheets and listen to mariachi music and the incessant barking of the neighbor's dog drifting over a cement fence crowned with shards of broken glass. Four bowls, six plates, one knife, five spoons. Tortillas, green salsa, mango tea. My son, Kip, plays wildlife games in the courtyard by moonlight. A hairclip belonging to Esmé, his sister, stands in for a lion, the twist tie from a bread bag has been fashioned into a radio tag. Toilet rolls for control towers, cereal boxes for airport hangers. Esmé performs ballet recitals in the midst of clothes-peg takeoffs: Gulliver in a Lilliputian airport.

The simplicity has been enchanting until this morning. Esmé crouches over the toilet, soiled and sweaty, screaming from the pain of diarrhea. There's not a drop of flowing water in the house. I hurry to the water heater and search for the on/off switch. Blanca, our landlady, had carefully demonstrated how to rapidly flick it back and forth when we first toured the house, but I must confess my Spanish was too poor to understand when and why I would do this. I try it now. Flick. Flick. Flick. When you only have one solution, you use it regardless of the problem.

"Find me a wrench," Andy bellows from the backside of a closet, where the water main is placed. My husband has lowered his six-foot-six frame into a space more fitting for a Jack Russell terrier and is dueling with the fitting for the intake pipe. "There aren't any tools in the house," I call back, as I yank open drawers in search of possible substitutes. Paring knife. Clothespeg. Empty matchbook. *Nada.*

Hours later, Esmé fully recovered and happily chasing chickens in the street, our calls for help answered, we watch as an elderly plumber and his stout nephew vanish into the closet. They carry huge pliers, wrenches, and hammers. "Thank goodness," Andy mutters above the din of clanging metal. "It would have been awful if they could have fixed it without tools."

Tools are such an essential part of our daily existence that I often take them for granted. This is not the case with two-year-old boys, who freeze in their tracks at the sight of a passing backhoe or concrete mixer. But by the time they reach preschool, many kids can group power drills and hammers as "tools" just as easily as they can sort piles of plastic animals into "farm" or "zoo." Tools play a pivotal role not only in our ordinary lives but in defining what it means to be human. Louis Leakey originally argued that the origin of the genus *Homo* related directly to the development of toolmaking. Only we can make tools, use tools, and interact linguistically. Only we live in behaviorally complex communities.[1] Or so our parents were taught.

Jane Goodall began shaking the foundations of that notion in 1960 with her discovery of a chimpanzee *(Pan troglodytes)* who fashioned a blade of grass to fish for termites. It was the first recorded occurrence of tool manufacturing in nonhumans in the wild.[2] Almost fifty years later, the evidence of chimp tool use is staggering. An international team of researchers, examining more than a million hours of chimp observations collected from seven long-term study sites across equatorial Africa, describe extensive patterns of tool use and cultural variation. Each group of chimps has its own distinct ways of grooming, gathering ants, probing beehives for honey, and making threats.[3]

Recent studies reveal even greater surprises about the depth and inventiveness of chimps. Today you can log onto the Internet and watch research video footage of an adolescent female chimp hunting a small primate with a spear.[4] Reminiscent of the murderous shower scene in the Alfred Hitchcock movie *Psycho,* the chimp repeatedly stabs her weapon into the cavity of a tree to kill a bush baby sleeping within. It's the first documentation of a nonhuman primate fashioning a weapon for hunting.

We've known that chimps use tools for less than half a century, yet new evidence suggests they've been doing it for thousands of years. The astonishing discovery of chimp "hammers" used by Stone Age chimps to crack nuts 4,300 years ago appears in a 2006 issue of *Science.*[5] Says Julio Mercader, the man behind the discovery, and one of the few archaeologists in the world to study the material culture of great apes: "We used to think that culture and, above anything else, technology was the exclusive domain of humans, but this is not the case."[6]

In his captivating book *The Ape and the Sushi Master,* eminent primatologist Frans de Waal provides overwhelming evidence that chimpanzees have a remarkable ability to invent new customs and technologies and that they pass these on socially rather than genetically.[7] Yet even with chimps, a species with whom we share 98 percent of our genetic makeup,[8] the idea of nonhuman animals using tools and having cultures is difficult to study, tougher to prove, and for some, impossible to accept. But guess what? Now there's evidence that dolphins use tools too.

Bottlenose dolphins *(Tursiops truncatus)* are the undisputed darlings of cetacean research. They've been studied more than any other cetacean species, thanks to their small size, their worldwide distribution, and their propensity to frequent the same near-shore waters we do. If you visited us in La Paz, I would send you down to the harbor to see them. They're the most commonly sighted cetacean in the Gulf of California.

People have been watching bottlenose dolphins in these Mexican waters for centuries. Yet no one, to my knowledge, has ever seen one using a tool. Since dolphins have no hands or feet with which to clasp or grab, it's hard to imagine what a dolphin tool might look like. Perhaps that explains why the first person to see a dolphin using a tool didn't know what he was seeing. It happened far across the Pacific, in Shark Bay, Western Australia, in the mid-1980s. The person was a fisherman. He thought he had spotted a dolphin with a massive tumor on its snout. He mentioned it to Rachel Smolker, one of the founders of the Shark Bay Dolphin Research Project. But it was the early days in what would become one of the world's most extensive longitudinal studies of dolphin behavior, and the dolphin could not be found. Rachel finally spotted the animal a year later. The growth did not turn out to be cancerous. Indeed, it didn't even turn out to be a growth at all—it was a marine sponge. The dolphin was using it as a tool to fish.

These days more than a dozen dolphins living in Shark Bay regularly pluck sponges from the seabed, hold them over their beaks, and use them to flush out fish.[9] The sponges may also act as a protective glove so the dolphins don't get stung by the venomous spines of stonefish.[10] Young bottlenose daughters learn how to "sponge" from their moms. The little ones start displaying the behavior around their third birthdays, close to the age when they are weaned. Since 1988 Janet Mann, professor of biology and psychology at Georgetown University, and her colleagues have studied the relationships between eighty calves born to more than sixty bottlenose dolphin mothers using the waters of Shark Bay. About fourteen of these mothers use sponges. Genetic analysis reveals they are all descended from the same female—a "sponging Eve," as it were.[11]

Although the individuals who use sponge tools are related, the task itself can't be explained by genetics. There is no gene for sponging. Instead, this form of dolphin tool use is culturally transmitted from mothers to daughters. One dolphin using a tool is news. Other dolphins imitating the behavior and passing it on through the population is much bigger news. It's evidence that marine mammals teach each other to use tools in the wild; it's evidence of culture. "We define culture as a behavior that is acquired by imitation and passed on in a population," says Michael Krutzen, the scientist from the University of Zurich who leads the new research. "We think this behavior is an example of that. It's very exciting."

But why is it almost exclusively the females who engage in this behavior? "Cultural transmission tends to hinge time and time again upon the young adult female," Lori Marino, a researcher in Neuroscience and Behavioral Biology at Emory University, tells me. "I don't think there's an exception to this in the primate literature or in the dolphin literature. In every case where we know which individual began a particular tradition, it's always been a young female."

Lori points to the case of the spear-hunting chimpanzees and the now celebrated example of Japanese snow monkeys who decided it was a good idea to wash their sweet potatoes in hot water before eating them. "It's the females that start things," she tells me. "In terms of tool manufacturing, use, and innovation, it seems like females are right up there."

Adrienne Zihlman, an anthropologist at the University of California at Santa Cruz, agrees. "Females are the teachers," she says, noting that juvenile chimps in the spear-hunting population in Senegal were repeatedly seen watching their mothers make and hunt with spears. "Females are efficient and innovative, they are problem solvers, they are curious."[12]

I think of mothers and their young as I chase Kip and Esmé along Balandra, a smooth, flat, sandy giant of a beach about fifteen miles north of La Paz. It twirls along a rocky coastline and licks the roots of mangroves in a gentle mix of dune and sea. The Gulf of California is famous for its tides. And on this crystal-blue day, the sea is so shallow the children make believe they are lobsters and sea otters and underwater photographers, their heads high and dry, small fingers and toes firmly planted on the bottom. The sea nourishes the insatiable creativity of child's play, and in the hours ahead, they will find sea pens and baby shrimp and porcupine fish so spiky you'd squeal just to see one. They will be buried alive, live in a fancy castle, tag tuna for research expeditions, and fashion jewelry from shells. Children are such masters of imaginative play, so willing to fashion a "this" from a "that," that it's easy to understand our long history of human tool use and innovation.

And now we know that other species share these talents. As I bend to lift Esmé high into the air, I realize we are dancing both figuratively and literally, on the edge of a sea of discoveries yet to come. More than two-thirds of the earth is covered by water. Bottlenose dolphins are but one of eighty-four different species of cetaceans. What other spectacular examples of tool use and exotic cultures lie hidden beneath the surface? The ocean is not the monocultural equivalent of suburban McMansion sprawl. It is a cultural mecca—an undersea amalgam of New York, London, Paris, Hong Kong, Delhi, and beyond.

7

Mirror, Mirror on the Wall,
Who's the Smartest of Them All?

The other morning Esmé raced into the bathroom, climbed up onto a stool, gazed into the mirror, and announced in a very disgruntled voice, "I look like a girl!" Apparently she had her heart set on seeing a boy gazing back at her. It was only after she'd gone off to rummage in her big brother's dresser and come back wearing a basketball singlet and baggy red jeans that her reflection produced a smile. She climbed back off the stool and proceeded to spend the next ten minutes eating breakfast as "Alex," her favorite big boy alter ego. Later that day, I found myself in front of another mirror; the three-way fluorescently lit variety that lurks behind the curtains in ladies' changing rooms. As I reluctantly engaged in the dreaded ritual of trying on swimsuits, I found myself in much the same position as Esmé: "Is that really what I look like? Can that possibly be *me?*"

Who do you see when you look in the mirror? Our human obsession with mirror gazing dates back at least eight thousand years to Turkey's Çatal Hüyük, the oldest and largest Neolithic city in the world. It was there that archaeologists discovered the earliest mirrors made of polished obsidian, a natural black glass created during volcanic eruptions.[1] Today an entire generation of carseat-bound babies occupy themselves by playing with the person in the mirror attached to the back of the passenger seat. By age two, most of these little ones will be fully aware that the handsome stranger smiling back at them is no stranger at all. The ability to recognize and be aware of oneself is among the most fundamental issues in psychology, from both a developmental and an evolutionary perspective. It is a defining human trait, shared only with a very few of our fellow primates, the great apes (go-rillas, chimpanzees, and orangutans). Or so it was believed, until Lori Marino and her colleague Diana Reiss of the Osborn Laboratory of Marine Sciences at the New York Aquarium put a mirror into the dolphin exhibit at the New York Aquarium.

Lori spends a lot of time thinking about mirrors and the big-brained in-dividuals who gaze into them. She holds faculty and research positions with Emory University's Neuroscience and Behavioral Biology Program and the Smithsonian Institution's National Museum of Natural History. She did her doctoral thesis on brain-behavior relationships in cetaceans and primates with Gordon Gallup Jr., a world authority in primate cognition. Forty years ago Gordon conducted the original mirror-test experiments revealing that chimpanzees are self-aware. The underlying idea of the experiment will be familiar to you if you've ever been told you have a piece of parsley lodged in your teeth. You can't wait to get to a mirror so you can see it and discreetly remove it. You're aware as you tilt your head to get a closer look at your re-flection that the green glop trapped between your left incisors is in your mouth. If the same thing happened to your cat, however, kitty wouldn't have a clue whose mouth she was looking at.

"You can put certain animals in front of mirrors forever, and they never learn that that's them," explains Lori. "Many pets and more exotic species will play with or respond aggressively to a mirror image as if it was another animal. They understand reflecting surfaces, but they don't have that level of cognitive awareness of their own being that you need to use a mirror to investigate your own body. A chimpanzee will take about a half an hour, and she's got it down."

Lori and Diana applied the mirror test to two bottlenose dolphins in 1998.[2] They first established a control by marking the dolphins with a marker filled with water rather than ink. After several repetitions, they put temporary black ink, which the animals could see only in a mirror, on parts of the dolphins' bodies. They then repeated the test by water-marking them once again. In every one of the sixteen videotaped tests, each dolphin swam directly to the mirror to investigate the place where it had been marked, often twisting and turning to get a better look at the proper spot.

Lori marvels at the extraordinary speed with which bottlenose dolphins grasp ideas. "I'm always struck by how ahead of me and everybody else they are. They seem to form hypotheses about what you want, and they seem to get impatient when you don't know what they're trying to convey. You almost feel like you hope you're smart enough for them. They can be very engaged and very quick."

In the early 1990s a number of zoos and aquariums began making a concerted effort to develop varied training regimes to provide more intellectually enriching environments for the cetaceans in their care. In those days I used to carpool to work at the Vancouver Aquarium with my friend Brian, a trainer of killer whales. "I just can't keep up with them," Brian would lament as we sat waiting in the early morning rain to cross the Lions Gate Bridge, thinking up games and activities for the whales. "No matter how many creative ideas I come up with, the whales are always three steps ahead."

My husband, Andy, a marine mammal specialist who now manages the Sea Otter Research and Conservation Program at the Monterey Bay Aquarium, would understand. For years he worked with dolphins and whales at Sea World in San Diego. "When you play ball with a dog, the dog brings the ball back," he tells me. "When you play ball with a dolphin, it invents a new game." He smiles as he describes the natural capacity of dolphins to innovate. "They'll dribble the ball on the water with their rostrums or hold it underwater and let it shoot up. They'll stop and blow bubble rings and roll the ball in those. They'll swim right up and look at you, and you just know they're in the mood for mischief."

For me one of the most captivating examples of the parallels that exist between humans and dolphins comes from a study conducted by Vincent Janik, an animal communication specialist with the University of St. Andrew's Gatty Marine Research Unit in Scotland.[3] Bottlenose dolphins, he says, are among the quickest learners of new sounds. As infants they develop their own "names," or signature whistles, which they use throughout their lives. Other dolphins in their group repeat these whistles back to them when they are socializing. In other words, they call one another by name. They share this enchanting custom with only one other species: us. "I think it is a very exciting discovery," Dr. Janik told the BBC News, "because it means that these animals have evolved the same abilities as humans."[4]

Just how intelligent are dolphins? Since the 1960s scientists have conducted controlled experimental research on bottlenose dolphins in the domains of memory, conceptual processes, vocal and motor mimicry, behavioral innovation, "language" understanding, mental representation, and self-awareness. Dolphins, it turns out, have insight into their own feelings and thoughts as well as into what they know and what they don't know. Their short-term and long-term memory for visual, auditory, and multimodal information, as well as for abstract concepts, is exceptional. In 2003 David Smith, a psychologist at the University of Buffalo, and his colleagues came out with a study that placed humans, dolphins, and monkeys in situations in which they were forced to perform cognitive tasks.[5] The trials were sometimes really difficult and sometimes fairly easy. The idea behind this method is to optimize rewards. Just like on a game show, you need to think about how certain you are of specific answers and not give them when it would be too costly. "That study is really phenomenal," says Lori, "because all you have to do is look at the graphs. You can superimpose one graph on top of the other, so whatever the three species are doing, they're doing the same thing. When put in that circumstance, the response patterns of humans and dolphins and monkeys are identical!"

The cognition of dolphins—how they think, know, and remember—so closely parallels that of great apes and humans that Louis Herman, a psychologist at the University of Hawaii at Manoa, has coined the term *cognitive cousins*. Yet the most intriguing implications of these studies may not be that dolphins can do so much, but that they do it with brains that are so structurally different from our own. According to Lori, the human brain and the dolphin brain represent two alternate ways in which large brains and complex intelligence are built.[6] "Dolphin brains are strikingly different from primate brains, and yet somehow they can accomplish the same sorts of things cognitively that a chimpanzee or a human can do," she says. "There is more than one way to evolve an intelligent species."

Lori has an encouraging, friendly manner and a knack for saying profound things in simple ways. "Big brains are not unique to humans," she tells me, and I believe her. I know that she has an intimate knowledge of her subject. Lori has read everything written about dolphin brains back to the 1800s. She examines the brains of dolphins and whales that strand and die on beaches. And she uses noninvasive imaging techniques to study living cetacean brains and the craniums of their fossilized ancestors.

"If I just looked at the brain of a dolphin, I would be fascinated by how elaborate it is and how different it is from ours," she explains. "If I just looked at the behavior of a dolphin, that would be fascinating too," Lori explains. "But you get something by merging the two and comparing them, and that's the message, that you can have a very different brain but as long as it is an elaborate brain, you can emerge with very sophisticated cognitive capacities. So the human way is not the only way to be smart. Or the great ape way is not the only way to be smart. There are other routes the brain has taken evolutionarily to get to the same place."

I realize, as Lori is speaking, that what she is saying is true. But I also realize that somewhere along the way I've learned that you're not supposed to say so. It violates a fundamental notion that we've all grown up with, the idea that nature involves lower and higher organisms and that we're at the top. We're special. Philosophers call it the Great Chain of Being.[7]

It reminds me of a conversation I had with pilot Sandy Lanham. Sandy was flying over Bahía Magdalena (Mag Bay), one of the gray whale calving lagoons, with John Ruthven, a producer with the wonderful BBC film series *Blue Planet*. Suddenly she looked down and spotted a line of gray whales bodysurfing. They were riding an outgoing tidal bore. "There's a calf trying to do it, but it slips off and starts swimming furiously to try to get back on," she tells me. As they watched, John turned to her and asked if she thought the whales were playing. "I have been so heavily criticized by whale researchers when I've told them I think whales play," she explains, "that I just wimped out and said, 'I think they're enjoying it.'"

"Do you realize it takes a double leap to immediately say they're not playing?" John replied. "The natural starting point is to say that they're playing. Do you think it is the arrogance of religion that makes us deny it?"

Lori understands the scenario well. "I think for religious reasons it is safer to believe that you're somehow in a special place above all the other animals," she explains. "But the Great Chain of Being is a philosophy that basically has no evidence from the scientific community. None at all."

"When I look at the brain of a dolphin or the brain of a honeybee or the brain of an aardvark or the brain of a person, they're all different because we're different species," Lori says. "But there's nothing about the human brain that is qualitatively different from any other brain. It's made of the same stuff. It's how the stuff is put together." Lori is sensitive to the dangers of being anthropomorphic, and she's conservative in her claims. Yet she also thinks it's possible to be so afraid of being anthropomorphic that one becomes almost ridiculous about it. "If a chimpanzee does something that's exactly like what a human does under exactly the same conditions, I don't think you're being anthropomorphic," she tells me. "I think you have a reason to say the task is probably being done in the same way. Evolution works by just tweaking things that already exist." She cites the example of killer whales showing their babies how to kill seals. "The mother brings the child to the shallows and kills a seal right in front of the child and provides an example, a model," she says. "The child tries it, and there's correction by the mother. If you looked at an orca brain, you would be humbled."

Evidence for intelligence in whales can be found not only in their behaviors, and the structure of their brains, but also in the composition of their cells. In 2006 a team of scientists at Mount Sinai School of Medicine in New York announced the discovery of spindle neurons in the brains of humpback whales.[8] These specialized brain cells had previously only been found in humans and our closest relatives, the great apes. They occur in parts of the human brain that are thought to be responsible for our social organization, empathy, speech, intuition about the feelings of others, and rapid "gut" reactions and are credited with allowing us to feel love and to suffer emotionally. "It's absolutely clear to me that these are extremely intelligent animals," says Patrick Hof, who codiscovered the whale spindle cells.[9]

The lives of dolphins and whales are extraordinarily complex. These animals live in highly diverse environments that require them to engage in a wide range of sophisticated behaviors to survive. Just the sort of conditions, Lori believes, in which intelligence evolves. "We are starting to see that there is a strong relationship between certain kinds of ecologies and brain size and cognitive abilities, not just in cetaceans but across the board. In primates, in birds, in some carnivores," she says.

"If you look at those birds or those primates or those dolphins or those carnivores that have the largest brains, they are also the ones that live in the most complex demanding social environments. They are also the ones that live in the most unpredictable physical environments. I think there's something about living in an unpredictable environment that is correlated with having to be very behaviorally flexible and intelligent."

Mirror, mirror on the wall. Who's the smartest of them all? It depends on the context. It depends on the environment. It depends on our openness to looking beyond our deeply ingrained narratives. I make a silent vow and put the Great Chain of Being to rest.

8

Building Nets from Bubbles and
Other Mysterious Humpback Whale Talents

Humpback whales blow bubbles ranging in size from small enough to get under the shell of a shrimp to as large as a party-sized pizza. But not all the bubbles come out of their mouths. Like all whales, humpbacks breathe through an opening called a blowhole on the backs of their heads. By tightening the muscles around their blowhole, whales change the shape of the opening and the size of the bubbles that come out.

Fred Sharpe knows more about humpback bubble blowing than anyone. The first time I met him, in the late 1980s, he was leaning over the edge of a large circular "bubble tank" in a laboratory at Simon Fraser University in British Columbia, trying not to topple under the weight of a full-size replica of a humpback whale flipper. Fred had seen groups of humpback whales in southeast Alaska blowing bubbles, and he suspected that the bubbles helped the whales catch fish. But because the whales blow their bubbles five stories beneath the surface, and then thunder to the surface, crisscrossing underwater with up to fifteen other whales at a time, he couldn't find a safe way to observe what was happening. The bubble tank enabled Fred to test how fish react to the types of bubbles humpbacks blow.

Today Fred is the executive director and principal investigator of the Alaska Whale Foundation. With nearly two decades of observations and experiments (both in the lab and at sea) behind him, he speaks with confidence: "Bubble net feeding is a form of tool-using behavior. The humpbacks are perhaps unique, with humans, in that it's a form of communal tool use. One individual will employ a net for the entire group." When a group of humpbacks gets together to feed, one whale takes the lead. While other whales in the group herd the prey (nearly always herring) to the area by making hunting sounds, the leader dives down beneath the school of fish and releases a stream of rising effervescent bubbles from its blowhole.

Next time you're pressed for conversation at a wine tasting, wet your finger and run it around the top of your wine glass. According to Fred, it makes a sound much like the fishing call of a humpback whale. The fish move up to escape the sound, and the bubbles rise with them, forming a tube-shaped net that keeps the fish from swimming away. "The rest of the pod shows a high degree of coordination and restraint," says Fred. "The bubble net has to mature. There's a point where it's optimized for prey capture." At that moment, the whales chase the prey up through the vertical tunnel of bubbles like space-bound school buses, filling their enormous mouths with the tons of fish corralled in a bubble net big enough to encircle a four-story house.

When he began his research, Fred expected to find that all humpbacks make and use bubble nets the same way. Instead, he discovered that bubble-netting is more like a team sport in formation and that the plays called by the captain change depending on what's happening in the game. Bubble nets vary in size and shape depending on which humpback whale is blowing the bubbles, how many whales are involved in the hunt, and what kinds of fish they are hunting.

Researchers and whale watchers identify individual humpback whales by the shape and pattern of their tail flukes (or lobes). These photo-identification catalogs act as yearbooks, enabling researchers engaged in many different studies to determine who's who and, when they've had the opportunity to take a clear tail shot, to contribute updated photos of existing individuals and new calves. Fred spends so much time listening to the feeding calls of individual humpback whales that he can also identify some of them by their sounds. Recently he successfully attached a crittercam tag (an underwater video camera hardy enough to be worn by a fifty-five-ton, fifty-foot-long lunging humpback whale) to "Melancholy," a known vocalizer. The camera provided Fred with a whale's-eye view of underwater bubble-netting. The video confirmed what he had long suspected. Whales in the feeding groups are task specialists. One individual focused on producing bubbles, while Melancholy, the primary vocalizer, was more focused on herding and on getting prey into the bubble net.

"It is just like on a football team," explains Fred. "There are certain quarterbacks and certain wide receivers that really click. They know each other's nuances and talents, and they make a great team." Fred has discovered that individual whales choose who they want to bubble-net with, often partnering with the same individuals for the entire summer, or from year to year. These social bonds can last for decades. "Ones that hang together are most likely individuals who have some sort of reciprocity. They're each gaining from their act," Fred says. "Both males and females can be leaders in these groups. These bonded individuals seem to have a higher rate of feeding and a lower rate of aborted bubble nets, suggesting that they are working together in a pretty efficient fashion. The groups of whales re-form each year in specific areas or work in discrete neighborhoods."

What's particularly exciting about these long-term bonded relationships is that the whales are not related to one another. "It is intensely humanlike," Fred exclaims. "I mean, enduring social bonds between nonrelatives, engaging in task specialization, and a form of collaborative tool use. We really don't have any better examples, other than humans."

I ask Fred how Melancholy got his name, and he smiles. "It's for the beautiful, highly modulated, flowing, enduring pulses to his calls. Kind of tears your heart out when you hear it." But Fred would also be the first person to challenge the idea that all is harmonious in the world of whales:

> When two feeding groups join up, often you'll get this big chorus of grunting. When you get these groups joining, it necessitates a change in what positions the individuals adopt. Imagine that you've been a quarterback, but now you're going to be relegated to being tight end. Not everybody's always happy

about that. We tend to view these as highly cooperative ventures, but with the crittercam, we could actually see the leader, Melancholy, get rammed by another whale while they were underwater. It happened a couple of times. It seems humpbacks are probably more like humans in that there are always some diverse perspectives around the office.

Fred finds humpbacks reminiscent of humans in other ways too. Like us, humpbacks seem to have a "grass is always greener" perspective. If groups are foraging in the same bay, and one group appears to be having a high rate of feeding success, you'll get a lot of defectors over to that group. As soon as that group's luck runs out, they'll race back toward other more productive groups.

The individuality of humpback whales impresses Fred. "You definitely see a huge range in the predilections of feeding preference and dietary choices of individuals. I've seen individuals that seem to be foraging on krill when other whales are moving all around and constantly switching. It is really hard to generalize, just as it is with people. There are as many strategies out there for exploitation of prey as there are individuals in the population."

Fred's descriptions of humpbacks in the southeastern Alaskan feeding grounds reminds me of kids on the playground at recess. They seek favorite friends with whom to play basketball, while keeping an eye out for more exciting groups or goings-on. When the class bell rings, they go their separate ways. Humpbacks also head off alone when the summer feeding season ends, albeit for more grown-up reasons. "There is no evidence that these enduring associations persist beyond seasonal habitats," Fred explains. Even in the summer feeding grounds, well-established partners will feed off by themselves on krill early and late in the season. Fred believes that these groups mainly feed in the summer when herring schools are abundant and predictable and the whales can find each other and form teams. The associations probably break off when the whales head south to breed near Hawaii because the males are competing for females. As Fred puts it, "Alaska is the galley, and Hawaii is the bedroom."

Humpbacks that use the waters around the tip of Baja and up into the Gulf of California as their "bedrooms" each winter are not part of Fred's study group. These individuals head to California or as far north as the central Gulf of Alaska to feed. Another population of humpbacks breeds farther south, three hundred miles beyond the tip of the Baja Peninsula, off Socorro in the Revillagigedo Islands. These animals migrate east to feed in an area stretching from Kodiak to the Bering Sea and the eastern Aleutian Islands. "If you look at humpback whales off Brazil versus humpback whales off Socorro, Mexico, versus humpbacks off California's Farallon Islands, versus Hawaii, they behave quite differently," explains Christopher Clark, the world's leading expert on whale acoustics. "Yes, the males sing . . . but they arrange themselves like primates do. The whole social system can be very different, depending on the context in which they're residing."

The complexity of these differences is coming into focus, thanks to an ambitious humpback research project, called SPLASH, spanning Mexico, Hawaii, Canada, Russia, Japan, and the United States. Jay Barlow, head of the Coastal Marine Mammal Program at the Southwest Fisheries Science Center of NOAA Fisheries, sits on the steering committee of this international cooperative effort both to understand the population structure of humpback whales across the North Pacific and to assess the status of, trends among, and potential human impacts on this population.

Jay and his colleagues hope to have collected biopsy samples from 5,000 individual humpback whales by the time the project's three winters and two summers worth of data has been analyzed. "There are several pieces of information we can take home from skin and a little piece of blubber," he tells me. "From the skin we can do the genetics, and we can look for the heavy metals. From the blubber we can look for the organochlorine pollutants, as well as the fatty acids, which gives us a measure of what they're eating. And we can even extract enough hormones from the blubber to tell whether the animal is pregnant. We can easily tell sex from the DNA content of the skin as well."

The science behind such specialized analyses is intriguing. Robots extract the DNA from the skin at the Southwest Fisheries Science Center in La Jolla, California. Once the DNA has been extracted and purified, it's flown to Scott Baker's conservation genetics lab at the University of Aukland. When I ask Jay about the challenges of transporting genetic material from an endangered species (the current designation of humpback whales) across international borders, he nods. "If it's actually the tissue from the animal, it's a tremendous hassle," he answers. "We *can* send tissue samples from one CITES (Convention on the International Trade in Endangered Species)- certified institution like ours to another, such as Scott's. But we try to simplify it by extracting the DNA and then making replica copies of the DNA and sending the replicas."

Scott is all too familiar with the importance of restricting the transport of threatened whale species across borders. He's been testing the identity of whale meat sold in Japanese fish markets for more than a decade. In 2004 he uncovered the first DNA evidence of illegal whaling in southern hemisphere waters by matching DNA-tested whale meat from the Japanese retail market with that of endangered sei whales *(Balaenoptera borealis)* from the southern hemisphere stock.[1] He and his colleagues in the lab have discovered that a number of other protected species—humpbacks, Asian North Pacific gray whales, Bryde's whales, and fin whales—are also routinely sold in the market. Scott's research adds fuel to the argument that Japan's whaling for scientific purposes creates a market for the illegal killing of whales, including endangered and threatened species.[2]

Genetic samples are just one form of data SPLASH is assembling. Another is photo IDs. When Jay tells me that SPLASH is collecting more than ten thousand unique photo IDs of individual humpbacks, I am eager to learn how they match them. The ability to pair photos of the same whale in both its northern feeding grounds and its southern calving grounds is critical to piecing together the puzzle of where different humpback populations travel. "We're seeing movement patterns that we've never had an opportunity to see before," Jay tells me. "Matches between islands in the far western Pacific, Japan and Okinawa and the Philippines, matching up to Russia and the Western Aleutian Islands. We figured that is the logical place that they would be going, but we had no real evidence up until now."

I assume that with such an enormous volume of images, SPLASH must be using some form of computerized matching technology. But it turns out that the tried-and-true method of having humans do the matching remains more accurate. If you were to peek into the offices of Cascadia Research in Olympia, Washington, you'd find a dedicated group of biologists sorting through digital images of humpback whale tails. "They're just powering through, matching individual photographs," Jay says with obvious appreciation. "There's a big culling process, in the beginning, to limit the photos to only good-quality ones, so if you saw another photograph of the same animal, you could match it. But still, it is an enormous task, and each season it grows exponentially. I'm glad we're done after three years. I'm not sure we could handle any more."

Scientists used to think there were no impediments to humpbacks swimming all the way across the Pacific. Originally each of the large whale species listed as endangered was listed as just one worldwide population. The picture emerging from the SPLASH data analysis presents another view. "The more we look, with genetics or with sound, the more finely populations seem to be structured in the oceans," Jay explains. "When it comes to taking individual populations off the endangered species list, there's no doubt that we're going to be doing it on a population-by-population basis."

That happy day may be swift in coming. "I fully expect that we're going to find that the total population of humpbacks in the north Pacific is over 10,000 individuals," says Jay. "Perhaps pushing 20,000 animals. That's fantastic news. It really means a major recovery from where they were when whaling was ended. It's difficult to consider a whale as truly endangered when you go up to a place like Alaska and you can park your ship twenty miles out to sea and can see humpback blows in every direction as far as you can see."

Humpback whales remain wonderfully enigmatic to Fred. When I ask him to share something about humpbacks that would surprise people, he is quick to answer. "I heard a really wild thing," he tells me, a bizarre observation made by his colleague Volker Deecke, research associate with the Marine Mammal Research Unit at the University of British Columbia. For some time researchers have noticed that humpbacks appear to be rubberneckers. "When killer whales are attacking either humpbacks or other animals, such as sea lions, the humpbacks are attracted to these sources of sounds and to the excited animals. Humpbacks will actually come over, almost blunder into the middle of these killer whale attacks." Amazingly, Volker observed humpback whales imitating the killer whales' predatory behavior. "You know the way killer whales ram and bat their prey with their tails?" Fred asks. "He's seen the humpbacks come in and slap the seals with their flippers, the seals that are being attacked by the killer whales."

Episodes like these clearly delight Fred. "You know, they are such hardcore cultural manipulators, such copiers. There is so much to these animals. You think you've got them figured out, and then you see this sort of mimicking, copying behavior, playing behavior. It's extremely rich." Fred shakes his head in amazement: "We still don't even know why the hell they breach. Obviously a lot of it appears to be for fun. But with so many of those basic things—breaching, pec slapping, making social sounds—it's very likely that it is just for individual amusement and for play, because so often it appears to be unpatterned and so individually specific. It's hard to attach a definite adaptive value and utility to it. A lot of these things suggest that humpbacks are a lot like us."

9

Do Baby Sperm Whales
Suck Milk through Their Noses?

Jeremy Bowen, the Middle East editor for the BBC, is talking on the car radio as I navigate the morning rodeo of cars vying for the coveted curbside parking spot outside the local *jardín de niños* (kindergarten) near our home in La Paz. I double-park, struggle to help Kip and Esmé across the busy traffic lane, and watch as they disappear through the gates into a sea of dark-haired children. The contrast between the playful shrieks of the children and Jeremy's description of the horrors he has witnessed while reporting on more than a dozen wars is so poignant that I instinctively reach to turn off the radio. It is at that moment that he speaks of his own child, how, in May 2000, he discarded the cloak of invincibility he had worn through every tough assignment and realized his life was hanging from a fragile thread. He clearly remembers the date of this transformation: it was the day his first child was born.

It seems odd that we can recognize the profound impact of birth on our own lives yet rarely mention it when discussing other species. We speak of whales, not whale mothers or fathers or babies or grandmothers. Killer whales live in arguably the most closely knit societies on earth. In some populations, both males and females remain with their mothers their whole lives, lives that for females may span eight decades. Sperm whale females do the same thing, raising their children in the company of their mothers, sisters, aunts, and close friends. Yet we recount their birth stories, indeed all aspects of their lives, in dry clinical terms and reduce their meaning to scientific data points.

Perhaps we do so because it's hard to think of a cold, gray ocean seething with large predatory sharks as a nursery or of a one-ton newborn sperm whale as a vulnerable infant. True, this mondo-baby can swim within a few hours of birth, but it can't remain underwater for very long. That's a problem for a calf that's completely dependent on a deep-diving mother. Mother sperm whales dive quickly, descending and ascending at a rate of about 213 feet each minute. They forage at depth, 1,640 feet or so beneath the surface, for half an hour before coming back up for air. Each dive lasts thirty to forty-five minutes, although sometimes the whales will stay underwater for well over an hour. Ten minutes of rest at the surface, and then it's back down for another long feeding dive. So who watches the baby while mom's away? Shane Gero, a graduate student in Hal Whitehead's lab at Dalhousie University in Canada, hopes to find the answer.

Unlike most whale researchers, who study whales from boats, Shane literally submerges himself in the sperm whales' environment, donning a snorkel and sliding in with a fifteen-ton female and her calf. "It's a pretty surreal experience," he tells me, fresh from the high of swimming for forty days in a row with a known pod of seven sperm whales off Dominica in the Caribbean. "The first time I did it, I was petrified."

Nicknamed the Group of Seven, a subtle nod to the famous Canadian landscape art movement, these seven individuals comprise a permanent social unit of grandmothers, mothers, female friends, and their children. They're the best-studied group of sperm whales in the world. Researchers recognize each by sight and have collected DNA samples from all of them. By staying with the whales as they moved through their range and recording who spent time with the little one, Shane identified one primary babysitter. Now he's waiting for preliminary genetic samples to confirm his suspicion that it's dear old granny. "She is definitely the largest individual," Shane says, "and based on the number of scars on her skin, she's old."

"Allocare," or care given by a nonparent, is pretty rare. Southern resident killer whales in British Columbia and bottlenose dolphins in Shark Bay, Australia, are the only other cetaceans known to do it. But within these populations, fascinating examples exist.

For some sperm whales, allocare extends far beyond simply keeping watch. If a group is without a calf, the whales will often synchronize their dives. You'd find them all down deep clicking and feeding for forty minutes. But if the group has a calf, the other whales stagger their dives. Adults will pass the calf off to one another. Females living in the middle of the North Atlantic (in the Sargasso Sea) not only babysit but also nurse each others' calves. As Shane puts it, "We found two different systems of allocare in sperm whales, one in the Sargasso and one in the Caribbean. The main difference is that it looks like the calves are suckling from multiple females in the Sargasso and from only the one mom, the one we're presuming is the mom, in the Caribbean."

Our own history of what the La Leche League calls "cross-nursing" extends back to 2,000 B.C. Shane's discussion of cross-nursing among sperm whales, and the exciting discovery of cultural differences between groups of whales that practice it, generated a lot of interest at the 2005 biennial conference on the Biology of Marine Mammals. But what really caused a stir was Shane's description of his next set of slides, the ones during which he talks about sperm whale calves nursing through their blowholes.[1]

Hang on a second. I can almost guess what you're thinking. Is this guy nuts? Before you hop to that conclusion, take a minute to consider the context. Shane is the only person studying sperm whales nursing from an underwater vantage point. "Being in the water with these animals is a very invasive way of getting data," he tells me. "Thirty years ago Hal Whitehead and Jonathan Gordon got in the water and said, OK, that's suckling. So people thought, now we don't have to get in the water with them and bother them so much." Shane decided to snorkel with the whales to distinguish between when calves were suckling and when they weren't. "A whale calf nurses by rolling onto its side and putting its mouth on the nipple," he explains. The trouble is, none of the sperm whale calf nursing sessions he saw looked like that. "What we saw every time was blowhole to nipple."

Shane is a conservative and thorough young researcher. He's quick to caution that the majority of these data comes from just one calf in the Group of Seven. Yet he followed that calf intensively for weeks, and he has video and photographs to prove what he saw. "It came to the point where I had to conclude that either this animal is in severe jeopardy, which it isn't, or that it's nursing," he says. "I put it in my talk because I wanted other researchers to give me feedback, negative or positive. If we are able to make a hypothesis as seemingly outrageous as this one, it shows how little we know about this behavior and that this work is needed." Shane's got a great point. Regardless of the whale species, researchers define suckling behavior based on what they see at the surface. "It's the same for humpback whales and other animals. We've made basic assumptions about them based on old research, but no one's really gone in to confirm them," he explains.

You can't help but admire a graduate student who's willing to shake the foundations of his field with solid, albeit very preliminary, data. The scientists at the conference felt the same way, awarding Shane the winner of the predoctoral spoken presentation award and flooding him with ideas. "The anatomists basically said it's possible, though none wanted to be the first to say that it's happening," he says with a smile. "We've nixed the idea of suckling at too deep a depth. A calf might dive and suckle at fifty meters or something, so we aren't able to see it, but both the behaviorists and the anatomists are sort of skeptical of that based on increased water pressure at depth. A calf could be suckling just at night, but that seems unlikely. Everyone I've talked to, particularly Janet Mann, an associate professor at Georgetown University who does the work on the dolphins, thinks that twelve-hour lag is just way too long."

Shane will return to Dominica in the hopes of finding and snorkeling with the Group of Seven again. If he can't with that particular group, he'll settle for any sperm whales, what he admiringly calls "ecologically dominant, massive, unbelievable animals" that are perfect for their environment. "We can define behaviors and give them definitions, give them numbers and quantify and do statistics, but ultimately we're studying an interaction between two individuals of a population. You know, a mother and her baby."

When I ask if I can join him on his next trip, he politely refuses. "I think it's an experience that I am extremely lucky to have, but unfortunately, I don't think it should be encouraged." He doesn't think swim-with whale programs are good for the animals. He does it solely to develop knowledge that he hopes will lead to sperm whale conservation. But for now, he's on a mission. "We want to settle this whole suckling thing," he says.

10

Deep Culture

Many years ago I joined Wade Davis, an ethnobotanist studying the pharmacological properties of rainforest plants, on a group expedition to Borneo. We traveled by longboat down rivers choked with logs destined for the Japanese forestry market searching for the Penan, one of the few remaining nomadic peoples of the rainforest. Their homeland in the Malaysian state of Sarawak is undergoing one of the highest rates of deforestation on the planet, and the goal of the expedition was to learn more about their traditional uses of plants and to draw attention to the political plight of these displaced indigenous people. I remember vividly the experience of watching a young boy hunting forest pig with poison blow darts. He was so silent, so skillful, that he would appear and disappear among the tree trunks as effortlessly as smoke. Could I ever really understand his perception of the environment, the depths of a culture so different from my own, I wondered, even with the benefits of an interpreter and shared language?

I find myself contemplating a similar question while sitting on a boat on a windless afternoon in the Gulf of California. I'm watching the bushy blows of sperm whales shooting in and out of view with equal grace, the clouds of glistening mist the only evidence of the elusive multiton beasts traveling below. A sperm whale's blowhole is located to the left of its forehead. Its blows thus project forward and to the side at an angle quite different from those of other whale species. Sperm whales too are nomads, and their cultures are even harder to observe than those of forest people. Mature males are three times the mass of the females, and the difference in their habitats is dramatic. Adult males are cold-water creatures, venturing as far as the ice edge at both poles. Juveniles and their fifteen-ton female elders live in the tropics and subtropics and may travel about twenty-two thousand miles a year within a vast home range measuring the length and four times the width of the state of California. "Sperm whales are hard to study," Hal Whitehead, the world's authority on sperm whales, says in a classic understatement that reveals his British roots. "They are constantly moving, and most of their behavior is deep underwater where you can't see what is happening."

These are not exactly the ideal parameters in which to study a foreign culture. Yet culture is exactly what Hal is interested in studying. He is widely recognized as the man who gave the "C-word" credibility with respect to cetacean research. The fact that he chooses to examine something as elusive as culture in arguably one of the most elusive animals on earth just makes him all the more impressive, like a pioneer of an extreme sport.

The analogy is a fitting one. Sperm whales, according to Hal, are animals of extremes. They are one of the largest animals on earth. The spermaceti organ is a strange thing: the world's most powerful natural sonar system. It makes incredibly strong clicks that go out into the world, bounce off things, and bring back information. Also in the head of the sperm whale is the largest brain on earth by weight (seventeen pounds).[1] The sperm whale is the deepest mammalian diver, going down at least sixty-five hundred feet, and probably farther. For those of you who enjoy the *Guinness Book of World Records,* you can add "world's longest intestine" and "world's longest muscle" to the list of sperm whale attributes.

No stranger himself to pushing limits, Hal encountered his first whale while sailing solo along the coast of Nova Scotia in 1974. Fresh from Cambridge University, with a degree in pure mathematics, he had no intention of studying cetaceans. But it wasn't long before he was smitten. In 1981 he and his colleague Jonathan Gordon pioneered the study of living sperm whales using a thirty-three-foot sailboat. Before then almost all sperm whale research was wrapped up with the whaling industry. Some scientists believed the animals were too elusive to be studied alive, a sentiment still expressed by the head of the Japanese Whale Research Institute and used as their rationale behind resuming the killing of sperm whales in the North Pacific in 2000.

In the early 1990s Hal spent a sabbatical sailing around the South Pacific with his wife, marine biologist Linda Weilgart, and their two young children. His daughter Stephanie was just ten months old when they left. They recorded sperm whale sounds and took photo-ID shots, amassing the beginnings of what has become an incredible catalog of about three thousand individuals. Each sperm whale is identified by a unique pattern of marks and scars along the trailing edge of its flukes. Hal and Linda also collected whale poop by racing over whenever they saw a whale leave a brown patch at the water surface and gathering what they could. Feces contain important information about what a sperm whale eats. Male sperm whales, according to Hal, are messy eaters, and their tendency to leave behind bits of squid beak provided important dietary information. Whale "dandruff," or bits of sloughed-off skin, was another treasure, a source of DNA for working out genetic relatedness.

Hal still has the same thirty-three-foot sailboat, but these days his battalion also includes an array of three-foot-long "minititanics." Sperm whales spend so much of their lives diving at depth, they are difficult to see. But thanks to their amazing sonar system, they are easy to hear. "All our recordings and pretty much everyone else's recordings of sperm whale sounds have been made with one or two hydrophones, underwater microphones, which give you just the sound that was made and maybe which bearing it was made on," Hal explains. "But if you record in an array, with a series of hydrophones, you can tell exactly where each sound is. The problem is that sperm whales are always moving or almost always moving, and the whale has to be pretty much in the middle of the array for the array to work. So we've developed a series of model unmanned boats. When the whales are behaving appropriately, we'll send out the model boats in an array around the sperm whales, and each boat has a hydrophone, and a GPS. It's fun. We have been testing it out, and it seems to work."

If the idea of listening to whales strikes your fancy you can buy a basic portable hydrophone (underwater microphone) to dangle off the end of your kayak for about the price of an iPod.

Hal and his graduate students use their system to listen to and follow groups of female sperm whales day and night, for up to a week at a time. The sounds Hal is most interested in are "codas," the patterned series of clicks that sperm whales use for communication. These repertoires remain stable for years. Hal likens them to a kind of Morse code. The sperm whale social units in the Galapagos he studied could be classified into three clans: the "regular" clans that have codas regularly spaced. CLICK CLICK CLICK. The "plus-one" units that make an additional click, almost like a Canadian "eh." CLICK CLICK CLICK__CLICK. And finally "unit T," which makes a totally different set of codas.

Whales with the same codas have a lot more in common than just sounds. They share distributions and behaviors. Regular units, for example, were found close to the Galapagos Islands and seemed to do well when it was relatively cold. In contrast, the plus-one units were found farther from the islands, some twenty miles away. They did relatively better when El Niños happened and everything warmed up. "When we found that there were, in each area, different clans producing different vocalizations, we then went to the other things that we can measure, such as movement patterns and microdistributions—or micro for sperm whales, which are on the scale of tens of kilometers," Hal explains. "More recently we've been looking directly at diet and reproductive success, how frequently they have babies, for instance, of the different clans."

The evidence points toward an intriguing conclusion. Sperm whales live in multicultural societies. Female sperm whales are cultural animals. "A young female sperm whale learns a whole lot from her mother and the other females in her social unit, including the dialect and the way they move around and probably a range of other things," says Hal. "In the South Pacific there are four or five clans who range over two thousand to ten thousand kilometers. Each clan contains tens of thousands of females. Members of the same clan are not necessarily related. From a zoologist's perspective, this is weird. The closest you get to this is *us*."

Sperm whales aren't the only multicultural society in the sea. Killer whales also form distinct cultural groups. Though they are found in the Gulf of California, they are not well studied in that region. Farther north, however, off the west coast of British Columbia and Alaska, killer whales are the best studied and perhaps most watched whales in the world. Killer whales in the Pacific Northwest are currently all considered the same species, yet they operate in three culturally distinct populations: fish-eating "residents," mammal-eating "transients," and a third, poorly described type known as "offshores" that inhabit the outer part of the continental shelf. Residents, transients, and offshore killer whales are strikingly different from one another in terms of what they eat, how they behave, the structure of their social groups, and their geographic range. Lance Barrett-Lennard, a research scientist at the Vancouver Aquarium Marine Science Centre, collects skin biopsies from photo-identified resident and transient killer whales to analyze their DNA. "What we've learned from DNA evidence is that these groups are genetically very distinct from each another," says Lance. "There's no interbreeding. They overlap in range. They look pretty similar to each other. They do different things, and they haven't intermated for at least hundreds, and probably thousands, of years."[2] Lance believes resident and transient killer whales are so genetically isolated that they are in the process of developing into separate species.[3]

The discovery of such profound cultural differences among killer whales in the Pacific Northwest and among sperm whales in the Galapagos and off Chile entices scientists to look for cetacean cultures in other places. "We want to look at sperm whale social structure in different areas in order to look at cultural patterns," Hal tells me. The Gulf of California tops his list of places to look, and it's easy to understand why. "Sperm whales are in the gulf all year round," Nathalie Jaquet, senior scientist at the Provincetown Center for Coastal Studies in Massachusetts, tells me. "There are quite a lot of them, about a thousand individuals. Most of the groups have at least one calf and sometimes two or three calves." Over the past decade Nathalie and

Diane Gendron and a number of graduate students have been tracking sperm whales in the gulf and recording their codas. "We have at least two clans," says Diane, "and there could be two different clans at the same time." Both Diane and Nathalie are quick to caution that their findings are preliminary. "We only have about five hundred codas instead of the ten thousand or fifteen thousand that Hal used for his study," Nathalie says. But the findings are exciting nonetheless. "If there are two separate clans or separate social groups or social megagroups," explains Diane, "they might have different adaptations to predators; maybe one will feed on different types of food, maybe not. How do they live together? It's all very exciting and very important for conservation."

I remember a paleontologist once telling me that his was a profession of delayed gratification. On average it takes seven years of full-time work to prepare a single fossilized dinosaur skeleton. Delayed gratification, I am coming to appreciate, could well be the watchword for sperm whale research too. "Sperm whales are puzzling," Nathalie says with a resigned smile. "It's often impossible even to find them. You can't look at a place and say whether it's a good one for sperm whales. It seems like you always have sperm whales where you don't expect them." Nathalie knew that in the Galapagos sperm whales often turn up where the slope of the seafloor is very steep and where there is lots of underwater relief. But when she surveyed areas with a similar ocean floor topography in the Gulf of California, she came up empty-handed. And then she went up to San Pedro Mártir. Located midway between the peninsula of Baja California and mainland Mexico, San Pedro Mártir is the most isolated island of the Gulf of California. "You have these absolutely flat undersea areas about seven hundred meters deep," she says. "Full of sperm whales."

If Nathalie dreams of finding ways to predict sperm whale behavior, Doc White dreams of the million-dollar sperm whale photograph that he will retire on. "The first time I got in the water with sperm whales was in the Sea of Cortez," Doc tells me. "They were about a quarter or a half mile away, swimming toward me. I dove to about twenty-five feet and was just kind of hanging out waiting, when all of a sudden I look down and there was a sperm whale swimming upside down looking at me. Later I swam with a sperm whale that was lounging on the surface with his mouth open. The teeth in the upper jaw are not erupted, but the teeth in the lower jaw are. Some people think that's how they hear. Better to perceive noises through that jaw. But when one's swimming at you with its jaw wide open, that's not the first thing that comes to mind!"

Although Doc has taken hundreds of sperm whale photographs, the prize he truly covets is one of these leviathans with a giant squid. He even knows someone who claims to have seen exactly that off San Benedicto Island, about two hundred and fifty miles south of Los Cabos, Mexico. As fate would have it, it was a friend's mom, and she didn't have a camera. She was sitting on the back of a boat knitting. She heard a splash, looked up, and beheld a sperm whale breaching. "It took her a moment to realize the whale had a giant squid," says Doc.

I can almost see Nathalie rolling her eyes. "People think that sperm whales feed mainly on giant squid, and that is likely not true." If giant squids were a big part of a sperm whale's diet, she argues, they would have shown up in the whaling records or in more recent studies of feces from living animals. Instead, researchers find lots of other species in whale excrement. Sperm whales have an eclectic diet that varies depending on where they live. Dozens of fish species, particularly sharks and rays, form an important part of the diet of sperm whales living off New Zealand, the North Atlantic, and the North Pacific. A wide variety of squid species, however, remains the mainstay of most sperm whale diets. These squid vary from the size of a pea to longer than an alligator. Rather than tucking into three square meals a day, sperm whales forage almost continuously day and night, nibbling on relatively small and not very nutritious squids in much the same manner as you and I consume popcorn.

Nathalie blames *National Geographic* magazine for planting the indelible image of a huge squid wrapped in battle around a sperm whale. Even the largest giant squid (at about eleven hundred pounds) is only about 5 percent of the weight of a female sperm whale—and a far smaller percent of the weight of males, which are their principal predators. Not much of a contest. "It's also unlikely that they stun their prey with sound," Nathalie adds, crushing another dearly held bit of sperm whale lore. With squid, in fact, it's the other way around. Just as they close in on the squid, they grow silent. "If they were stunning it we would hear a big noise," says Nathalie. "I don't know how the idea got started, but it really doesn't seem to make any sense." Hal takes a softer tack, stating that the possibility that sperm whales use sound to stun their prey remains open, although it seems unlikely that it is a common feature of sperm whale foraging. He believes that the sperm whale's nose functions principally as incredibly powerful sonar. In his marvelous book *Sperm Whales* he writes: "In an environment where sperm whales compete with other sperm whales in a demanding scramble for mesopelagic squid, she who finds the most squid will prosper and multiply. Any attribute that allows its bearer to detect squid more effectively will be favored."[4]

It is squid that drew Nathalie to the Gulf of California, the jumbo, or Humboldt, squid *(Dosidicus gigas)*. These predatory animals weigh almost as much as I do and can be three feet taller. The gulf is one of the few places where there is overlap between sperm whales and a fishery for squid, and it's the only place anyone has been able to look at both sperm whales and their prey at the same time. The squid fishery in Santa Rosalía is one of Mexico's largest fisheries, yet the techniques have changed very little since the 1940s. Fishermen still set out in small pangas, drop lures overboard, and pull up their catch by hand. During fishing season, as many as three hundred pangas will go out in the evening and return the same night with up to twenty-two hundred pounds of squid per boat.

William (Bill) Gilly, professor of biology at Hopkins Marine Station of Stanford University, has been studying Humboldt squid in the gulf for decades. A few years ago, he and Nathalie bumped into each other quite by accident at a marina in the Gulf of California. "We didn't know they were putting these satellite tags on squid," says Nathalie, "and they didn't know we were putting them on whales." This serendipitous meeting marked the beginning of a rare opportunity to study the behavior of both predator and prey at depth.

"How sperm whales feed at depth is still a big mystery," Nathalie explains. "They are slow and awkward, and they have this tiny little jaw." Jumbo squid, on the other hand, are fast and muscular. So how do the whales catch them? "What we see in the Gulf of California, what we see at depth," Nathalie says, "is that sperm whales feed on the squid in the daylight hours when the squid are in the oxygen-minimum layer. The squid are going to be very sluggish and very slow and not at their best because there is so little oxygen. The sperm whales don't care because they get their oxygen at the surface."

It's a good theory, but it is in the process of getting debunked. The oxygen minimum layer (OML) is a midwater layer of oxygen-depleted water. It results from the microbial metabolism of sinking organic material generated by high surface productivity. In the region around Santa Rosalía, the OML typically begins around the six hundred- to one thousand-foot range during the day. That's the same depth at which sperm whales are often recorded and the predominant daytime location of Humboldt squid.[5] At dusk, the squid rise to the surface, a fact well understood by fishermen who attract them by shining bright lights on the water surface at night.

Data collected from the tagged animals reveal some incredible surprises that challenge the idea of sperm whales casually sucking up stuperous squid. Rather than remaining either at the surface or at depth, the squid frequently zoom between these two zones. Such activity suggests they may be actively hunting. "We can actually see the rates at which a squid's going up and down during the day and night," reports Bill. "Some are quite fast—from fifty meters down to four hundred meters in less than ten minutes."[6] Some squid spend up to six hours in the OML, where the oxygen concentration is less than 5 percent of surface values. Yet another type of tag demonstrates that a single squid can travel a hundred miles in just two weeks.[7] Because of these amazing feats, Louis Zeidberg, a postdoctoral student in Bill's lab, calls Humboldt squid "the athletes of the cephalopod world." As he puts it, "they act like there's eight milliliters of oxygen in the water when there's less than one."[8]

For now the mystery of how sperm whales feed on such energetic squid remains hidden in the deep. The tendency of squid species to form large groups, however, may provide part of the answer. Sperm whales may increase their hunting success by targeting highly concentrated groups of squid. Off the Galapagos Islands, groups of sperm whales with high feeding success zigzagged back and forth over areas about twelve miles across, suggesting patches of prey of this scale, whereas off northern Chile the patches seemed larger, nearly forty miles across. Different squid species spawn at different times and in different ways, which could further explain the wide dietary range of many sperm whales. For the most part, Hal thinks sperm whales get by on easy-to-catch smaller squid that are not only slow swimmers but also glow in the dark, thanks to their bioluminescence.

Sperm whales, perhaps more than any other whale, epitomize the relationship between people and whales. During the eighteenth and nineteenth centuries, whalers made the sperm whale the world's most commercially important animal, eventually driving them to commercial extinction and guaranteeing the species an indelible role in human culture. The discovery of fossil fuels actually saved them. These days sperm whales are poised at the brink of another deadly competition with us, a competition for food. Consider this: 350,000 sperm whales exist today. Add up how much they eat, and the numbers are staggering. Hal estimates that sperm whales take 80 to 100 *million* metric tons out of the ocean each year. That's a bit more than we manage to take out of the ocean with all our fisheries worldwide. This means that the sperm whales are using a huge ecosystem about which we know very little. No worries, you might think. Apart from the occasional plate of calamari, you probably don't eat much squid. But our global human culture is changing ocean systems so fast that we need to start asking what we will eat in the very near future. A 2006 issue of *Science* projects the collapse of all species of wild seafood currently being fished by the year 2050.[9]

Our ability not only to extract single species but also to impact entire ecosystems has frightening consequences. Biologist Juan Pablo Gallo Reynoso has been studying the ecology of the Gulf of California for the past twenty-five years. "They finished with the sharks in the year 2000," he says. "There were fourteen species of shark overexploited. Because there are no sharks left, the squid boom. Now squid are eating everything." Juan Pablo shows me a photo of the contents of a squid. Its belly is full of sardine paste. Squid grow so fast and are so precocious, Juan Pablo worries that dolphins won't be able to compete with them for sardines. "We seldom see big groups of dolphins like we saw years ago," he explains. "That is also why there are more sperm whales here now. They are here because there are more squid. It is all shifting."

Humboldt squid are apparently a recent addition to the gulf. Historically common off the coast of South America, they moved into the Gulf of California in the 1970s. Humboldt squid live less than two years, but they reproduce and grow quickly. Evidence suggests that 10 million squid may be living just within a twenty-five-square-mile area outside Santa Rosalía. More than 120,000 metric tons of squid were caught in 1997, most of it shipped to Asia, where it is a staple food. Today Humboldt squid have expanded their range north into Monterey Bay, where they are eating their way through the Pacific hake (or whiting) fishery. Although the invasive range expansion coincides with rising ocean temperatures associated with El Niños and global climate change, Louis Zeidberg, the coauthor of a 2007 report on the issue, thinks the growth in numbers may have more to do with the overfishing of their main predators in the Pacific, such as tuna, marlin, and sailfish.[10]

Currently there is only a small overlap between what sperm whales eat and what we do. But that situation could change quickly, especially when you consider both the rate at which we are depleting conventional seafood choices and the capacity of big-brained cultural animals like humans and sperm whales to adapt to new situations.

In 2006 an international symposium of scientists, fishers, and fisheries managers gathered in British Columbia to discuss the growing problem of what to do with sperm whales and killer whales that are learning to raid fishing gear.[11] In Alaska, for instance, some sperm whales have learned how to remove sablefish and halibut from the hooks of long lines, an astonishingly nimble act when you consider the proportion of their enormous heads in relation to their tiny lower jaws. The situation is not unique to the Alaskan waters. Sperm whales have been reported taking fish from long lines in the North Pacific, North Atlantic, Subantarctic, and South Pacific Ocean basins. The problem of depredation by sperm whales and killer whales is growing around the world. The cause and the mechanism of this increased spread are closely related: less natural food available in the oceans for whales and the capacity of these intelligent animals to learn how to raid fishing lines from each other.

Killer whales are behaviorally conservative. They're slow to take advantage of new food opportunities. Hal illustrates this point with an example from the North Atlantic:

> Killer whales are pretty scarce in this region. When we do see them, we see them making a difficult living trying to catch large humpback whales or trying to catch stuff in the ice and so on. And yet on Sable Island, near Nova Scotia, year-round there are two hundred thousand seals that are easy for an animal like a killer whale. There used to be a killer whale group in the area which was destroyed by whaling. We don't know what they were eating, but it is quite likely that part of their diet was the Sable Island seals. And since that time the seal population has increased a couple of orders, it is now several hundred thousands. No one has ever seen killer whales eating them, yet I have seen killer whales a few hundred miles away. It looks as though the knowledge that there is this wonderful source of food hasn't gotten through.

The fascinating thing about highly developed social learners like killer whales and sperm whales, however, is that once one or more individuals decide to sample something new and become good at acquiring it, others learn quite quickly to do the same. The behavior spreads rapidly through the population. "One charismatic killer whale says, 'Wait a moment, there's a whole bunch of seals,' and the whole thing changes," says Hal. The trouble with fishing depredation is that a lot of fisheries operate in the whales' natural foraging areas. This both decreases the available food for the whales and increases their exposure and thus the likelihood that one of them might try to take advantage of the fish being hauled up on a line. It is much more effective to try to prevent this from happening in the first place than it is to try to change the whales' behavior once they've developed a taste for the catch.

The same is true of us. Now that we've developed a taste for overexploiting fish, wood, and fossil fuels, it sometimes feels impossible to turn back. I think of our insatiable demand for the earth's resources. I think of the young Penan boy hunting wild pigs in Sarawak and wonder what has become of him and his nomadic culture in the twenty years that passed since my visit. I Google "Penan" and discover that the members of the young boy's tribe are still locked in blockades against foreign timber companies.[12] Only a few hundred individuals continue to lead a truly nomadic life, the rest relocated to permanent settlements. The forest they are vying for, however, is much diminished. In the past forty-five years, more than 90 percent of the virgin jungle of Sarawak has been logged. Though we are slow to change our ways, locked into a cycle of overexploitation of forests, oceans, skies, we are quick to take advantage of new opportunities, opportunities that feed our ravenous appetites. "We may well develop the desire and technology to catch sperm whale food, and we have a remarkable record of overexploiting marine resources," Hal tells me. "What will happen to the sperm whale then?"

Blue whales feeding, Coronado Islands.

A blue whale opening its mouth to take in a school of krill is the biggest biomechanical event to happen on the planet. During peak feeding, a single individual may swallow four to six tons of the stuff each day. Photos by Doc White.

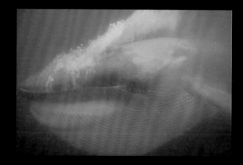

Sequence of a blue whale diving, Magdalena Bay.

When Doc White spotted this blue whale through his binoculars, he noticed what looked like a small pinhole through the left side of its fluke. When the whale came closer, he determined that the hole was actually the same diameter as his arm. In this series of images you can see the water pouring through the hole as the whale prepares to dive. It gives you a sense of the enormous size of this gorgeous fluke, which is probably twenty-five feet across. Photos by Doc White.

Blue whale calf breaching, north end of the Bay of La Paz.

Adult blue whales are rarely seen breaching, but calves, like this one, sometimes fly from the water at a forty-five-degree angle before splashing back down onto their stomachs or their sides. Photo by F. Gohier.

Fin whale blowing, Sea of Cortez near Espiritu Santo Island.

Like all baleen whales, this fin whale has a double blowhole. Depending on weather conditions, you can sometimes identify individual whale species by the shape of their "blows"—the moisture-laden cloud of air produced each time they exhale. Fin whales and blue whales have tall, straight blows. Gray whales have heart-shaped blows. Humpback blows are bushy and balloon-shaped. Sperm whale blows veer off to one side. Photo by Doc White.

Bubble netting, southeast Alaska.

Humpback whales in southeast Alaska deploy huge bubble nets around fish schools or krill swarms. Then they devour their prey in a spectacular communal lunge, as up to two dozen whales come rocketing up through the center of the bubble net. Photo by Doc White.

Humpback breach, Socorro Island.

This humpback was two miles away when Doc first saw it breach. He thought it would stop before he could get close enough to get a photo, but the whale kept breaching. Doc raced to his rubber boat and headed toward the whale. At this point it had breached seventeen times. He followed beside the humpback as it breached sixty more times in a row. He shot this photo near the end of this remarkable display, yet the whale was still leaping almost entirely clear of the water. Photo by Doc White.

Bottlenose dolphins leaping in the late afternoon light, Sea of Cortez, north of San Jose Island.

Like chimpanzees and spider monkeys, bottlenose dolphins travel in small groups that often change in composition. The tightest associations exist between pairs or trios of adult males. Photo by F. Gohier.

Two gray whales, San Ignacio Lagoon.

These gray whales are two of the approximately 20,000 individuals who return to the birthing and breeding lagoons along the Pacific coast of Baja each winter. These tranquil waters were once the primary killing fields for a historic population that scientists now believe numbered close to 96,000. Photo by Doc White.

Two adult blue whales, Sea of Cortez, about fifty nautical miles south of Loreto.

Blue whales typically socialize at a scale too large for humans to witness. When blue whales are seen together, it's usually in small groups of two or three adults, or in mother-calf pairs. Photo by F. Gohier.

Humpback mother and calf, Pulmo Reef, Sea of Cortez.

Mother humpback whales are legendary for their dedicated parenting. Unlike killer whales, however, they do not maintain a close social bond once the calf grows up. By the time a young humpback reaches sexual maturity (around five years of age), its association patterns with its mother are indistinguishable from its associations with other whales. Photo by Doc White.

Sperm whale mother and calf, Azores.

Whalers knew all too well how to use the strong maternal nature of female sperm whales to their own advantage. They would harpoon but not kill calves and then slaughter the females who stood by them. Photo by Doc White.

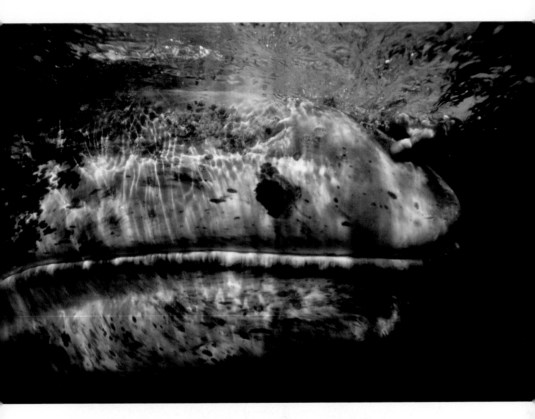

Gray whale and hand from boat, San Ignacio Lagoon; gray whale, San Ignacio Lagoon.

No one knows why some gray whales are friendly. Common wisdom among gray whale scientists is that these individuals are attracted by the sound of the outboard motors or the Jacuzzi-like sensation of its jets. Whatever the reason, the experience of having a whale hurry over to your boat and reach up to touch your hand is magical. Photos by Doc White.

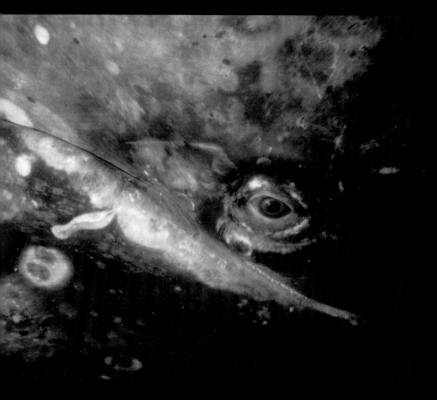

Gray whale eye, San Ignacio Lagoon.

To gaze into the eye of a whale is a rare and wondrous thing. Our challenge is to both revel in and commit to conserving the great complexity of the earth's oceans, whether or not the whales are watching us. Photo by Doc White.

What's the Use of Granny?

"Three years ago I had my feet in these stirrups for a completely different reason," I lament to my ob-gyn. Then I was delivering a baby. Today, according to this exam and the blood test results she holds in her hand, I am in menopause. I glance over at my preschooler doodling on her Etch A Sketch in the corner and am surprised by the power of a single-word diagnosis to make me feel suddenly and unspeakably old. If I were any other species of mammal, I remind myself in silent consolation, this announcement would be really bad news. It wouldn't just be about my identity. In most cases it would spell my death.

The end of fertility signals the end of life for most wild animals. A female's "value," at least in evolutionary terms, is to reproduce, and when the egg supply she was born with becomes depleted or impaired with age, she dies. As I unpeel myself from the paper covering the examination table and struggle back into the dignity provided by my jeans, for some reason my mind fixes on female sockeye salmon, their spent, bloated bodies scattered across the rocks of a British Columbian stream. Swimming an average of eighteen miles a day, these incredible mothers travel nearly two hundred miles inland from the Pacific Ocean to the stream where their own lives began. They arrive battered and torn from the trek, their bellies swollen with thousands of blood-orange eggs. The eggs are laid and fertilized, and after that, the females are lunch, literally, for the hundreds of eagles and bears that gather on the banks.

"The change," as my grandmother called it, is clearly a less heroic venture. But it's no less compelling. For much of medical history, we were the only known animal species in which females lived decades past the end of their reproductive years. Jared Diamond, author and professor of geography and physiology at the University of California, Los Angeles, calls menopause one of the most important features of human sexuality. "Along with the big brains and upright posture that every text of human evolution emphasizes," he writes, "I consider menopause to be among the biological traits essential for making us distinctively human—something qualitatively different from, and more than, an ape."[1]

Menopause may indeed separate us from gorillas and chimpanzees, but as so often happens in animal ecology, claims to exclusivity are bound to be refuted. One-quarter of all adult female short-finned pilot whales *(Globi-cephala macrorhynchus)* killed by whalers prove to be postmenopausal, judging by the condition of their ovaries.[2] Female pilot whales enter menopause at approximately forty years of age, and under ideal conditions they may live to see their sixtieth birthdays. Killer whales, sperm whales, and likely a few other species of cetaceans also enter menopause in their forties and routinely live on for several decades. Photographic evidence suggests that some individuals live for more than a hundred years, but like humans, probably only a small percentage. "These long-lived toothed whales have menopause just like humans do," Hal Whitehead confirms. "Females stop reproducing at age forty and then live another forty years. Our theory is that it is what they know that is important. In these kinds of animals, in these kinds of societies, knowledge is hugely important."

The societies Hal is talking about are matrilineal, stable groups of related and, in the case of sperm whales, also unrelated, females. Elephants and killer whales are among the most famous examples, and it may not come as a surprise that they too experience menopause. Like us, female elephants, killer whales, and sperm whales remain with their mothers for decades, if not for their lifetimes, nursing for months and sometimes years and learning many of the details of daily life from their moms and their grannies. Menopause and matrilineal social systems go hand in hand for long-lived, highly intelligent animals that invest years of maternal care, leading Hal and his colleagues to argue that menopause is adaptive and results from the trade-off between continued reproduction and assisting kin. Scientists label the behavior "eusocial," in the sense that we spend a substantial part of our adult lives reproductively sterile and helping our close relatives—a familiar concept to many of us in the "sandwich generation," caring for little ones and elderly parents at the same time.

The value of an old whale granny or elephant granny, or me, for that matter, Hal claims, is our life experience. Our memories. The things we've learned that can benefit the other members of our families, or matrilines. As we age, so this evolutionary argument goes, we do more to increase the number of people bearing our genes by devoting ourselves to our existing children and our potential grandchildren than we would by assuming the very real health risks of having yet another child. Hal cites the example of sperm whales and super El Niños. "Super El Niños are now happening every ten years rather than every eighty," he says. "When an El Niño strikes, sperm whales move. During the super El Niño of 1982–83, whalers found the animals in completely different areas off of Peru than they normally do." Hal suspects that the oldest females knew where to go when conditions got really challenging. Elephants in Namibia do much the same thing. "When drought becomes extremely severe once in many decades, they know to make a long, long trek to places where there is still water," says Hal. "We believe it is that knowledge that is being kept."

In his writings about a traditional community of subsistence farmers and fishers living on Rennell Island, one of the small Solomon Islands, Diamond draws a human parallel. "They ranked wild fruits in three categories":[3] those that people never eat; those that people regularly eat; and those that people eat only during famine, such as they did following a destructive cyclone that hit the island around 1910. Islanders survived by eating the fruits of wild plants, species that were normally not eaten. Knowledge about which plants were poisonous or how the poison could be removed by some technique of food preparation was passed along by aged survivors of the previous big cyclone.

Remembering the location or safety of food isn't the only kind of grand-motherly know-how that postmenopausal females share with their kin. The ability to remember if someone you haven't seen in a long time is a friend or a foe also turns out to be a remarkably important quality in ensuring that your genes get passed along from one generation to the next. Elephants can distinguish friends from strangers by the sounds of voices—both calls you or I could hear and lower-frequency rumbles that carry great distances.[4] A typical elephant group in Kenya encounters twenty-five other families, or about 175 elephants each year. Strangers might spark disputes or harass calves, so unfamiliar calls send a herd into a defensive huddle and a general state of distress. In contrast, a herd of familiar neighbors doesn't pose much of a risk. As incomprehensible as it seems to those of us who find it increasingly dif-ficult to remember people's names, the old adage "an elephant never forgets" appears to be true. The oldest female elephant in the herd is best at telling friends from strangers. The older the matriarch, the less her group reacts to other groups. "It's a great example of a group changing its response to a sit-uation because they were in the presence of an older, postreproductive fe-male," Shane Gero, who works with Hal, tells me. "It's like she's saying, Oh, I know them. You may not have met them, but I know they're OK."

The age of the matriarch also turns out to be the best indicator of the re-productive success of the herd as a whole. The oldest matriarchs know their environment better than younger leaders do and can lead their respective groups to better resources. Karen McComb, a behavioral ecologist at the University of Sussex and the lead researcher on the elephant study, believes the social sophistication of older elephants may also help boost their herds' successes.[5]

The savannah and the deep sea are completely different environments, yet so many aspects of African elephant social structure and life history closely match those of sperm whales, that it's intriguing to consider the selection pressures that might lead to this convergence. Both these species have low birthrates, prolonged parental care, and long life spans. They live in close-knit, stable, matrilineal units. They're also endowed with big bodies and big, intelligent brains. Perhaps most remarkable, they both possess truly unique and useful noses. Far back in the fossil record, tinier versions of sperm whales and elephants are found, with spermaceti organs and trunks seemingly fully formed—evidence, Hal argues, that these specialized noses evolved early on, giving these animals a feeding advantage over others. They came to dominate their ecological niches, and eventually, as other similar species became extinct, to evolve into the one-of-a-kind species they are today.

To understand the evolution of postmenopausal females and matrilineal units in sperm whale societies, however, you need to consider the ocean. "The ocean is naturally a pretty variable place over long timescales," says Hal. "Culture is a route to deal with that. If you have long-lived animals in a stable culture in a stable society, the oldest individuals give a lot to the younger ones just by what they know." Hal envisions ancestral female sperm whales as fairly solitary foragers that stuck quite close to the shore. It was only when these "presperms," as he likes to call them, moved out to the open ocean, with its widely dispersed and highly variable food sources, that these prehistoric relatives had to become constant travelers. There are no places to hide for animals that are almost continually on the move through the deep ocean. A permanent set of companions gives a female sperm whale better protection against predators, a greater ability to fight them off, and a babysitting system to help protect her calves. They also possess useful knowledge about where to go when conditions deteriorate. In a deep and watery world of endless variation, female sperm whales find refuge in the company of each another.

"To me the most interesting species are the large toothed whales with matrilinear social structures, like killer whales and sperm whales," says Hal. "Once you have this long-term bond between individuals, basically lifelong, then culture can really take off and have a major impact on genetic evolution, on conservation." He's begun adding other whale species—such as long-finned pilot whales *(Globicephala melas)*, short-finned pilot whales, false killer whales *(Pseudorca crassidens)*, and narwhals *(Monodon monoceros)*—that look like they have matrilinear social structures to his list of current and future research priorities. Hail the evolutionary power of aging grannies! Does anyone know how to get Hal anointed the patron saint of menopause?

Recognizing the importance of grannies is bittersweet. Understanding the value of their wisdom makes their loss that much harder to bear. "We'll probably never have an opportunity now to look at blue whales in their real context, because the density has been so destroyed," says Christopher Clark. "It's like some of Hal's descriptions of sperm whale populations where you take out all the large individuals. Or elephant populations. You've removed all the wisdom, and you've removed all the elders and all the larger animals, and you're never going to have the population structure the way it was through the vast majority of the evolution of the species."

Suddenly a train roars through my body. It's unstoppable and very unpleasant. I am unbearably hot and want to rip things off—scarves, jackets, turtlenecks. Beads of sweat on my forehead, between my breasts, and on the back of my neck turn into rivers. Flap. Flap. Flap. I fan the neck of my shirt, race to the freezer, yank open the door, and lay my head against a bag of frozen peas. In this moment I truly envy sperm whale matriarchs, plunging ever deeper in the cold, wide ocean. Then it's over. I mop up and start shivering.

12

Dolphin Snatchers

It's kind of like being mugged: you never see it coming. We are skipping across the surface of the southern Gulf of California, a flash of blue boat on a blue, blue sea, fourteen eager whale watchers and two experienced guides squinting through their polarized sunglasses into the white-hot glare of sunlight that bounces off the water. Nothing. Then *wham!* Hundreds of bodies shoot from the sea and spill alongside the bow of the boat, crowding so close we see the glint of their eyes. They know what they want, and they know how to get it. We are compelled to comply. The boatman revs the engine higher, and we're thrown back against the rail. Sandals slipping, cameras clicking, we struggle to capture the beauty of these bow-riding dolphin bandits.

I would wager that there are dolphins bow-riding on somebody's boat at any given moment of the day or night. Fishing boat, pleasure cruiser, ocean liner: if you're at the right speed, there's a good chance you'll have company. Pacific white-sided dolphins *(Lagenorhynchus obliquidens)* have even been spotted riding the pressure wave of an eighty-foot-long blue whale. I crane my head over the side to watch as a shadow of dark water forms deep beneath the boat. I feel more than hear the *zzzing* of sonar clicks. It is difficult to know the exact moment when the water ends and the glistening rocket of dolphin becomes airborne. It is only when it lands and veers sharply to the right to circle back underwater for another go that I realize the dolphin is really two dolphins. This mother and calf spotted dolphin *(Stenella attenuata)* are so closely synchronized, I can barely hear a pause between the "puh" and "puh" of their individual breaths.

Dolphin calves swim "kissing close" to their mothers for up to three years. Newborn dolphins can't control buoyancy well, and they tend to pop like corks at the surface. For the first hours of life in the sea, an infant dolphin rides high on its mother's flank, less than the length of your finger from her body. Soon the infant moves down to a more lateral position, its dorsal fin level with that of its mother. In this way the calf can ride in the slipstream alongside its mother with very few fluke movements of its own. Like experienced cyclists in a peloton, mother and calf adjust their positions in relation to one another as the little one grows. The calf will spend increasing amounts of time swimming under its mother's tail section, with its head lightly touching her tummy.

People have long wondered how young dolphins can swim fast enough to keep up with their speedy mothers. Slipstreaming helps a lot, but it doesn't fully explain the youngsters' speed. Daniel Weihs, an aerospace engineer at Technion, the Israel Institute of Technology in Haifa, believes that physics, not just love, keeps the pair tightly aligned. According to Daniel, the calves swim so close to mom they actually get sucked along.[1] You may not have thought about this phenomenon, known as Bernoulli's principle, since high school science class, but you experience it each time you take a shower. The rapid flow of the cascading water creates a low-pressure zone, sucking the shower curtain inward. Similarly, the movement of water around the speeding body of a mother dolphin tends to pull the calf sideways, in toward her flank. "The surprising thing is how strong the effect is," Daniel says. "The calf doesn't move its tail; the mother does, and both of them move. It's amazing." The force is so strong that mature dolphin cows have been known to "baby-snatch" other females' calves by sneaking up rapidly and "sucking" the babies toward them. "No one knows why," he tells me. "The assumption is that they're either a 'childless' female or one who lost her own calf prior to the snatching occurrence."

But what would cause such an attentive mother to lose a healthy calf in the first place? Some invariably fall prey to sharks or killer whales or accidents or disease. The more ominous villain in this tale, however, tears babies away from their mothers unintentionally. Like a magic spell in a fairy tale, the attraction between mother and child only works if the dolphins are swimming very close together. If someone or something drives the two apart, the positioning of the pair is broken, and the little one quickly falls behind. And that's what appears to be happening to mother-calf pairs of spotted and spinner dolphins *(Stenella longirostris)* off the mouth of the Gulf of California and farther west of Baja and Central America, a region collectively called the Eastern Tropical Pacific (ETP). Rather than playfully chasing boats for a bow ride, the dolphins are the ones being pursued. Rounded up by motor boats, trapped in purse-seine nets, and eventually, we hope, chased back out to sea, they are collateral damage for the yellowfin tuna *(Thunnus albacares)* fishery.

In a tuna-fishing chase, four to six speedboats operate as sheep dogs, herding the dolphins, their babies, and the tuna together, while the main boat encircles them with a one-mile long, four-hundred-foot-deep curtain of net. For reasons no one knows, tuna stay with the dolphins during this process, and thus tuna and dolphins are captured together. The chase tends to last between fifteen to thirty minutes, though some last up to an hour. Adult dolphins try to escape through a combination of bursts of speed, leaps out of the water, and brief periods of coasting. In a chase, the mother dolphins probably just run away and assume that the babies can just draft alongside.[2] Tragically, there is growing evidence that they can't. Data collected by official observers placed with the tuna fleet found that 75 to 95 percent of lactating female dolphins killed in the nets were not killed with a calf.[3] "Chases by fishing vessels can easily cause the loss of the mother-calf connection," Daniel says.[4] "If the mother really gets scared and starts moving really fast, the calf just can't keep up."

Some calves may be caught in the nets and eventually perish. Others probably escape, only to die soon afterward, without their mothers' milk and care to sustain them. The probability of death is highest for babies less than a year old,[5] yet none of these statistics appears on official industry records, which historically measure only the number of dolphins *observed killed* during sets. According to a key government report, "unobserved calf mortality potentially could be large, and continuing at the present time, if mother-calf separation occurs during the chase portion of the fishing operation."[6]

But I am getting ahead of my story. If you're like me, there's a can of "dolphin-safe" tuna sitting on your cupboard shelf. You buy it because your kids love tuna, everyone in your family cares about dolphins, and you have a general idea that it's the more responsible environmental choice. The idea that something labeled "dolphin-safe" could be caught in a manner that causes baby dolphins to lose their moms and die doesn't make sense. How could such a contradictory situation exist? And more important, what are tuna-eating, dolphin-loving people who don't want to spend a moment more staring in confusion at the plethora of tuna brands on the supermarket shelf to do about it? A quick cook's tour through the "tuna-dolphin issue," as it's euphemistically called in conservation circles, will help answer that question—I promise.

Fishing for yellowfin tuna didn't always involve dolphins. Mike McGittigan, the founder of SeaWatch, an environmental watchdog organization dedicated to the conservation of the Gulf of California, remembers bucolic days in the late 1950s when you could catch all the yellowfin tuna, or yellowtails, as he calls them, you wanted with a pole and line right off the beach. He describes fishing for tuna in Loreto, a beautiful coastal village in the southern Gulf of California. "The water would just boil with them for two or three weeks," he says. "They were feeding on sardines in the bay there." These happy memories were obliterated a few years later when Mexico ended up with the biggest tuna fleet in the world. "They brought in two big Campbell tuna seiners and ravaged twenty-four hundred tons of yellowtail from the same little bay," he says. The twin technological developments of synthetic netting that would not rot in tropical water and a hydraulically driven power-block to haul the net made it possible to deploy very large purse-seine nets around entire schools of tuna. "The fishermen were crazy," Mike exclaims. "The ice truck would come up full of brand-new Yamaha motors, because the guys were going out with these nets and catching so many yellowtails. Those were the rich days of killing the fish in the Sea of Cortez."

It was during those same days that U.S. fishermen first discovered that yellowfin tuna (or "chunk light," as it's often labeled on the can) school under large groups of spotted or spinner dolphins. Because dolphins, like all cetaceans, need to surface to breathe, they are far easier to spot than the fish hidden fifty or a hundred feet below. Find the dolphins, and you find the fish. "Dolphin sets" ("porpoise fishing," as the fishers call it) using purse-seine nets took off, rapidly replacing the slower and more targeted technique of catching individual tuna using a pole and line. Accidental entanglement and death of cetaceans is a global problem of enormous proportions, but that's not what we're talking about here. What makes this fishery truly unique is that fishers who engage in dolphin sets *intentionally* capture both tuna and dolphin together. More than 6 million dolphins have died in this fishery (the highest known death rate of dolphins for any fishery) since the practice began in the early 1960s. For comparison, the total number of whales of all species killed during commercial whaling in the twentieth century is about 2 million.[7]

Much has changed in the tuna-dolphin fishery over the past three decades. In response to public outcry, in 1972 Congress passed the Marine Mammal Protection Act, which prohibited U.S. fishing boats from dropping nets on dolphins, and then imposed the same standard on foreign imports in the 1980s. Since 1994 only "dolphin-safe" tuna—tuna caught *without* setting nets on dolphins—can be legally sold in the United States, the world's biggest market for the canned fish. By selling only "dolphin-safe" tuna, major U.S. tuna processors effectively banned imports from Mexico and other Latin American companies with large tuna fleets still engaged in dolphin sets. Mexico threatened action against the United States on the grounds that the U.S. dolphin-protection laws violate the free-trade rules of the World Trade Organization (WTO). Those nations that lost a huge part of their market owing to the higher standards, as well as other nations fishing in the ETP, joined together in the International Dolphin Conservation Program and implemented changes to both net construction and fishing techniques. Before pulling the net in, for example, the main boat turns and moves backward, drawing the net into a long channel. The speedboats head for this part of the net and hold it open so the dolphins can swim out. "If all goes well," states the official website of the Southwest Fisheries Science Center (the scientific research arm of NOAA's National Marine Fisheries Service, Southwest Region), "the dolphins are released alive, but the process requires skill by the captain and crew, proper operation of gear, and conducive wind and sea conditions. As with any complicated procedure at sea, things can go wrong, and when they do, dolphins may be killed."[8]

Dolphin deaths in tuna nets in the ETP have since declined by 99 percent in the international fishing fleet and have been virtually eliminated around the U.S. fleet.[9] Eventually, under heavy trade pressure and in order to avoid trade sanctions by the WTO, the United States agreed to ease embargoes and to improve market access for ETP tuna coming from Mexico and other Latin American fleets. Now here's the kicker. Under the proposed agreement, the term *dolphin-safe* takes on a dramatically different meaning. On December 31, 2002, the director of the U.S. National Marine Fisheries Service delivered a chilling New Year's Eve message that expanded the definition of "dolphin-safe" to include tuna caught when dolphins were trapped and released but not killed. Under the new rules, tuna harvested by purse-seine vessels using nets intentionally set on dolphins would be eligible for being labeled "dolphin-safe," even if dolphins are entrapped, as long as an onboard observer certifies that no dolphins were killed or seriously injured during the set in which the tuna were caught.

The Commerce Department justified the redefinition based on this claim: "The number of dolphin deaths has dropped dramatically from over 133,000 in 1986 to less than 2,000 dolphins in 1998. . . . We believe that continued low level of dolphin mortalities will allow depleted stocks to recover to healthy population abundance levels."[10] The trouble is, the numbers don't add up. Spotted and spinner dolphin populations *aren't* recovering. The northeastern offshore spotted dolphin population remains at just 20 percent, and eastern spinner dolphins at 35 percent of their prefishery levels. According to the Commerce Department's own scientific researchers, dolphin mortal-

ity since 1993 "has been a very small fraction of population size, so that recovery of the dolphin populations has been expected. By 2002, however, there
was no clear indication of a recovery for either northeastern offshore spotted or eastern spinner dolphins."

The Commerce Department's announcement horrified environmental
groups. Led by lawyers from the Earth Island Institute, the decision was chal-
lenged in court and overturned with a legal decision issued on August 9, 2004.
U.S. District Judge Thelton Henderson found that Commerce Secretary Don-
ald Evans not only failed to conduct the scientific research required to relax
existing tuna-labeling laws but also engaged in "a pattern of delay and inat-
tention" to build support for his position. "The record is replete with evidence that the secretary was influenced by policy concerns unrelated to the
best available scientific evidence," Henderson wrote in a strongly worded
fifty-one-page opinion. "This court has never, in its twenty-four years, re-
viewed a record of agency action that contained such a compelling portrait
of political meddling." The best scientific evidence, he writes, "indicates that
dolphin stocks are still severely depleted and are not recovering despite extremely low reported mortality rates . . . and that indirect effects from the
fishery can plausibly account for the lack of recovery."[11]

One "indirect effect" is the separation of mothers and calves. But the list
of possible consequences of chasing and entrapping dolphins doesn't end there.
Heart attacks, muscle damage, miscarriage, heat exhaustion, psychological
breakdown—reading through the literature on dolphin stress associated with
tuna nets is a grisly and depressing business, especially when you consider
how frequently individual dolphins are captured.[12] The more sets a dolphin
experiences, the less likely it is to survive. A northeastern offshore spotted
dolphin—and if she is a nursing mother, then her calf as well—may be chased
into purse-seine nets an average of 10.6 times and captured 3.2 times each
year. Other scientists estimate even higher rates of capture—essentially one
capture a week—for individual dolphins traveling in groups of a thousand
or more.[13] So little is known about how often groups of such enormous proportions form, or whether individual spotted or spinner dolphins remain in

one group or move fluidly through groups of different sizes, that the impact of the fishery on individual animals is impossible to calculate. As with earthquake victims, the enormity of the horror and loss is most poignantly felt through the stories of individuals and tabulated at the level of populations. "There is concern that fishing operations may be causing chronic stress in [spotted dolphins], affecting their populations," write researchers from the Veterinary Medicine Faculty of Universidad Nacional Autónoma de México, reporting on a series of chase, encirclement, and stress studies (CHESS). It is still unclear how dolphins cope with the chase, encirclement, capture, and backdown during a set, and as a result how this relates to the welfare of individuals.[14]

Little attention has been given to the effect of human-caused violence and environmental destruction on animal sociality, psyche, and emotions. The impact of the purse-seine fishing on individual dolphins, and by extension on dolphin societies, simply doesn't exist in an industrial fishing culture that interprets all life forms in the world's oceans in terms of "fishery stocks" and "fishery takes." Yet the implications of chronic stress on the incredibly strong bonds between mother and calf, and the social structure of dolphin societies, demands that it should.

Consider this. You and I may not know each other, but we probably know some of the same people. At least we know someone who knows someone who knows someone we have in common. "It's a small world," we say when a new acquaintance turns out to be linked to us through a chain of other people. It happens surprisingly often. Everyone, so the saying goes, is separated from one another by no more than six degrees of separation—by six friends or friends of friends. The six degrees of separation hypothesis originated in 1929 in a short story called "Chains" by the Hungarian writer Karinthy Frigyes. A range of social psychologists, mathematicians, and Internet gurus have been applying and confirming its validity ever since. The idea is famous for inspiring the play *Six Degrees of Separation* by John Guare and the subsequent 1993 movie starring Will Smith. You can even test the theory yourself by logging on to the website of the computer science department at the

University of Virginia and linking the relationships between any film star in the world and the actor Kevin Bacon, via the Oracle of Bacon website.

Now it turns out that dolphins are even more closely connected to one another than we are. David Lusseau, an ecologist at Dalhousie University, has spent more than a decade studying the social networks of dolphins to find out who knows whom and how often they meet. In the 130-strong community of bottlenose dolphins living off the east coast of Scotland, he discovered it takes an average of just 3.9 steps to link any two dolphins by the shortest possible route through mutual friends.[15] Lusseau previously found a small world in a New Zealand population of bottlenose dolphins with 3.4 steps between animals. Such tight social networks might facilitate the rapid spread of news, such as the location of a food source. "Adult females serve as hubs of information. They have a large number of associates," explains Alejandro Acevedo-Gutiérrez, an ecologist at Western Washington University. He believes dolphin communities could be very vulnerable to the loss of a few key individuals. The impact of 6 million dolphin deaths imposed by the tuna fishery on the cultural stability of dolphin populations in the ETP is impossible to fathom.

Trauma affects societies directly through an individual's experience and indirectly through the collapse of traditional social structures. The effects of violence persist long after the event. Studies of human genocide survivors indicate that trauma early in life has lasting effects on the brain and behavior. A December 2005 article in the *Washington Post* reports a drastic increase in post-traumatic stress disorder (PTSD) claims in the past five years. "Experts say the sharp increase does not begin to factor in the potential impact of the wars in Iraq and Afghanistan, because the increase is largely the result of Vietnam War vets seeking treatment decades after their combat experiences."[16]

Gay Bradshaw, a psychologist with the environmental sciences program at Oregon State University, studies human-caused breakdown of elephant communities in Africa and Asia. In 2005 she and a team of prominent wildlife researchers made international news with their announcement that elephants too suffer from PTSD. "Elephant society in Africa has been decimated by mass deaths and social breakdown from poaching, culls, and habitat loss," they write.[17] "From an estimated ten million elephants in the early 1900s, there are only half a million left today. Wild elephants are displaying symptoms associated with human PTSD: abnormal startle response, depression, unpredictable asocial behavior, and hyperaggression."

According to Bradshaw, elephants are renowned for their close relationships. Young elephants are reared in a matriarchal society, deeply embedded in complex layers of extended family. Culls and illegal poaching have fragmented these patterns of social attachment by eliminating the matriarchs and older female caregivers (what scientists call allomothers). Calves witnessing culls and those raised by young, inexperienced mothers are high-risk candidates for later disorders, including an inability to regulate stress-reactive aggressive states. Even the fetuses of young pregnant female elephants can be affected by prenatal stress during culls.

If our understanding of dolphin societies truly does lag a century behind our knowledge of land animals, as Bruce Mate claims, it will be too many years before the full impact of the yellowfin tuna fishery on dolphins is understood. For now the meaning of "dolphin-safe" continues to be debated in the courts, and no standardized system for labeling tuna exists.[18]

These are the thoughts that haunt me as I gaze at the "classic tuna" option on the Subway Sandwich menu. "Dolphin-safe tuna and creamy mayonnaise is lovingly blended together to make one of the world's favorite comfort foods," the sign reads. Trouble is, I no longer feel comfortable.

13

Friendly Mothers, Friendly Calves?

At sunset, it shimmers like a silver thread, snaking across the cactus-covered flatlands and up through the gently undulating hills. It is prettiest then. Low lighting covers many blemishes, and this is no less true for roadside litter. By day the endless pieces of broken glass and plastic bottles that line the Baja highway are a foe to be wary of: "Be careful where you step!" I shout when Kip hops out to pee behind the car door. "Don't walk outside without your shoes." Yet when the sun sinks low in the sky, we speed through the heat of the early evening on a ribbon of detritus transformed.

Whale watching takes place on water, but each adventure invariably begins and ends with long car rides to and from the launch sites. These road trips are a mixed blessing. I revel in the beauty of Baja's desert landscapes, where candelabra cacti and boojum trees the shape of upside-down carrots stand alert among the tranquil tangle of velvety brown boulders and dry creekbeds. Yet I find myself saddened by the all too familiar juxtaposition of nature and human debris. Earlier this year we drove another stretch of "wilderness" up the northern coast of Vancouver Island in search of killer whales in Telegraph Cove. Kip passed the final hours of the drive gazing at the thick green forest beyond the window, exclaiming over the bears, wolves, and foxes that were surely hunting within it. I felt torn between a desire to nurture his childhood sense of wonder and the harsh reality that this "forest" was nothing more than veneer, a narrow corridor left by the forestry companies to shield our view from the clear-cuts beyond.

I am not alone in feeling the tension between the beauty and the beast that is environmental degradation. This road trip that takes us along Baja's litter-strewn highways was inspired by the promise of wilderness, as it is for so many others. We come to see gray whales in their breeding and calving lagoons on Baja's Pacific Coast, but what we are really hungry for is a taste of nature unspoiled.

It is a five-hour drive from La Paz to the closest of the three major gray whale lagoons. We make games out of spying on goats and counting vultures and fill the drive with songs and stories and snacks. Yet no matter how cheerful our intent or attractive the destination, the hours of traveling in a hot car with little ones are marked by a rising crescendo of wails and complaints. We pass a cluster of wreaths by a car lying crushed and abandoned at a bend in the road, and I am thankful that the children can't see out of the windows in the backseat. There are no streetlights to mark the way when the sun eventually sets, and we navigate the last thirty miles or so along a trunk road in the pitch-dark. It's a mind-numbing mix of exhausted quarrelling punctuated by surges of adrenaline when an oncoming oil truck swerves into our lane in an attempt to pass on the single lane, and dozens of rabbits dart across the road in an apparent reenactment of the "white death" scene from *Watership Down*.

We are off the beaten track. Or so it seems. It's a somewhat intoxicating sensation that is further reinforced by our first moonlit glimpses of Puerto San Carlos along the shores of our destination, Bahía Magdalena. We smell a fish-processing building long before we pass it and pull into the only motel in town. Mexican friends in La Paz recommended that we ask for a whale watching guide, or *pangero*, named Fito, and we are giddy with anticipation as we enter the deserted open-air foyer and ring the bell on the desk. It takes several attempts before a friendly Spanish-speaking woman emerges from a room with a blaring television and we embark on the affable journey of trying to explain with our limited language skills our desire for both a room and Fito. Kip and Esmé abscond with several sweets from the woman's candy jar before we comprehend that the motel accepts only cash. We do not have enough. Since there is no ATM in town, Andy leaves the kids and me in our room and spends the next two hours making the round-trip back to the nearest place large enough to have one. The whereabouts of Fito are still unclear.

"Mag Bay," as the guidebooks call it, is a remarkable destination, one of the only places in the world where one can not only see but also *touch* whales. Yet there is little evidence of this exceptional status as I turn on the spigot and watch the kids swirl around on the broken tiled floor of the shower stall amid strands of hair left by the last guest. With its bare bulb hanging from the ceiling and threadbare comforter, this room is a far cry from the slick new luxury resorts rapidly multiplying along the shoreline of Cabo San Lucas to the south. It feels good to be away.

We fall asleep listening to the night sounds of dueling televisions in adjoining rooms—Doctor Phil on one side and moans from the perfect women and men having consensual casual sex in opulent LA houses on the Golden Channel. The next morning a sharp knock jolts us awake. It's Fito, and he gestures for us to hurry. We rouse the kids and dress them in two layers of dust-covered ice-cream-stained clothes to keep them warm on the boat. I grab homemade buns from a basket in the foyer and race to the parking lot, where Cesar and Daniel, a gay couple on holiday from Mexico City, are waiting. We hop into the car and follow Fito through streets of sand past rough-looking dogs to a house with a boat parked on a trailer in the driveway. Esmé is suddenly desperate for a bathroom. There are no public facilities. Cesar volunteers to act as translator, and within moments, Fito pulls out his cell phone and calls the wife of the boat owner from her driveway, and we are welcomed into her home by a licking machine masquerading as a tiny Chihuahua. We finish our business, bound back outside, and hop onto the boat. It's a short but jarring ride atop the trailer through the potholed center of town and down a boat ramp. I am sorry but not surprised to see the high tide line marked by a band of broken flip-flops, candy wrappers, condoms, Styrofoam cups, and plastic bags. We are shoved out to sea. We are whale watching. Or, to be more precise, we are watching the ocean in the hopes of seeing whales.

It is a whale watcher's dream. You spot a blow far off in the distance. You raise your binoculars to your eyes and zero in on the slow roll of back as it disappears beneath the surface. You anxiously scan the water, hoping to catch sight of the next breath, a faint cloud of mist from which you'll determine where the whale is heading. But this whale makes it easy. It comes *toward* you. Fast. Within moments, it fills the binoculars. You fumble to free yourself from the strap around your neck and scramble for your camera. The stench of foul breath almost knocks you back. A whale is *rubbing its head* on the side of *your boat.* You click and click and click your camera, cursing the time delay as another blast of warm whale breath envelops you in soft, oily, incredibly stinky rain.

One of the thrills of whale watching is its impossibility. You engage in the pursuit knowing full well that in most cases, you may not even glimpse your quarry. I remember a similar sense of being on a fool's pursuit when my sisters came home from church camp one summer when we were young girls and excitedly taught me how to beat the bushes for a "snipe hunt." We spent the summer in hot pursuit of snipes, conjuring exotic images of these fearsome creatures. The hunts were thrilling, but we saw not a one. It was not until I took a course at a bird sanctuary several years later, in a failed attempt to impress a guy, that I learned snipes were real birds. Chicken-sized, dull, mottled birds that live in swamps.

It's not that people don't know that whales are real; it's just that we rarely experience the full extent of their realness. Even whale researchers feel the divide. These animals spend more than 95 percent of their time completely underwater in a habitat foreign to humans. Bruce Mate puts it well: "We visit it for short periods of time with awkward equipment, we're never fully comfortable in it, and yet these mammal cousins do everything there."

The gray whales off Baja defy these norms. The mythical beast you've been pursuing switches roles and becomes the pursuer. "It's incredible," says whale researcher Mercedes Eugenia Guerrero Ruiz. "In the calving lagoons the gray

whales decide if they want to get closer to you." Francisco "Pachico" May-
oral, a small-scale coastal fisherman in San Ignacio Lagoon, was the first person to describe the experience of meeting a friendly gray whale. While he was
fishing for black sea bass in the winter of 1972, a gray whale popped its head
out of the water and started gently rubbing itself on the side of his boat. The
whale's behavior shocked Pachico, a man raised on tales of the aggressive nature of these "devil fish." For the next forty minutes the whale submerged and
reappeared at various positions around his boat, eventually moving in close
enough for Pachico to cautiously extend a fingertip to touch it. The whale
didn't move. It stayed to be patted. Nearly forty years later, Pachico and his
family all make their living as eco-tour guides, forming one of dozens of companies that now offer intimate encounters with Mexico's friendly gray whales.

The thing about friendly gray whales is that they don't look friendly. They're
huge and expressionless and covered with sharp-edged barnacles as hard as
cement. On closer inspection, you can see the orange glint of whale lice firmly
ensconced in the narrow valleys of skin not yet covered by barnacle armor.
There is no puppy wiggle, no gentle kitten pawing. These massive beasts lurch
from the water like enormous boulders trying to unfreeze from the spell of
a wicked enchantress. Yet they come to your boat. They rise to touch your
hand if you extend it. Friendly gray whales shove their massive bodies across
the air-water divide. "It is a very distinctive behavior that some animals exhibit and some don't," says Liz Alter, an evolutionary biologist and ecologist
at Hopkins Marine Station of Stanford University. "My colleague Sergio Flo-
res Ramirez was telling me that in San Ignacio Lagoon there was one whale
that had multiple propeller wounds on its rostrum, yet it would keep coming back. Season after season it would keep coming back after the boats."

No one knows why gray whales are friendly. Common wisdom among
gray whale scientists is that some individuals are attracted to the sound of
the outboard motors or the Jacuzzi-like sensation of its jets. Mercedes believes the attraction of whales to boats also depends on the skill of the *pangeros*. "Some *pangeros* are new in the area, and they are learning how to do
the speed changes, and maybe that won't give confidence to the mother," she

explains. "On the other hand, there are *pangeros* or fishermen who have been working there for all their lives, so they know how to approach the whales properly, how to make more sound with their boats."

Perhaps it's not so surprising that in whales, as in people, "teens" sometimes get a little carried away. "One time we were around a pair of gray whales that were frisky," Doc White says. "They started lifting the boat on their heads." Jorge Urban, coordinator of the Marine Mammal Research Program of the Autonomous University of Baja California Sur, tells me that young gray whales have also been known *not* to migrate to the northern feeding grounds. "Usually those whales are juveniles," he laughs. "They think that they can do anything. They don't want to do as all gray whales do!"

Liz uses molecular genetic techniques to shed light on the conservation, population biology, and systematics (the relationships among living things, both past and present) of marine organisms. She is eager to discover if gray whale friendliness is passed on genetically from mother to calf or whether calves could pick it up from an unrelated individual just by watching and learning. Apparently some calves are curious about the boats when they are just two days old, but the mothers prevent any interaction, often physically placing themselves between the panga and the babies. Indeed, in San Ignacio Lagoon, whale watchers are prevented from entering the northern part of the lagoon where females typically give birth. By the time the baby is a month old, however, the mothers seem to relax and may allow the little ones to approach.

To make sure her sampling technique is pleasant for the whales, Liz has come up with a novel way to collect bits of skin—a Brillo pad on a stick. When a friendly mom or calf swims over for a scratch, she'll treat the animal to a back rub or head rub. "You can get plenty of DNA that way," she assures me. "Usually folks feel that it's often superior to get a biopsy sample, because you can get some blubber that you can use for toxicological analysis and you get a lot more high-quality DNA. But certainly you can get plenty from sloughed skin itself."

The bigger challenge turns out to be finding skin to rub in the first place.

It's a creepy concept, but gray whale skin is literally alive with other creatures. At least three species of whale lice and one species of barnacle are permanent residents, the latter boring headfirst into the skin.[1] It's not uncommon for a gray whale to carry more than four hundred pounds of these tiny beasts. So attached are these small creatures to their carrier whale that scientists can identify individual gray whales by their distinctive patterns of cream-colored barnacle and whale lice encrustations and the blotchy scars these parasites leave behind.

Whereas barnacles spend their larval days free-floating along ocean currents passively awaiting a new whale to colonize, whale lice are passed on during the sex act or from mother to calf. Christopher Callahan, a young graduate student at Humboldt State University, recently removed and identified every one of the 316 parasites found on a dead baby gray whale calf that had washed ashore in Humboldt County, California. All the beasties were lice, and 95 percent of them were located near the mouth and throat grooves, clearly suggesting that colonization occurred when the calf was nursing.[2]

For Liz the phenomenon of friendliness is just one of the remarkable insights one can gain from genetic analysis. What she's really interested in, though, is what molecular genetic techniques can reveal about gray whale populations across grand time scales. How many gray whales existed before they were hunted by humans? What are the implications of current environmental shifts, such as climate change, for their future populations? "The history of a population is written in its DNA," she tells me. "Genetics has emerged as this really powerful tool to help us gain insight into what the oceans used to look like before large-scale human impact. Even in the sort of daily grind of lab work, when I think about it, it's always a little bit mind-blowing."

Gray whales make an ideal subject for this kind of work. In 1994 California gray whales became the first, and so far only, marine mammal to be taken *off* the U.S. Endangered Species List. "Gray whales are an amazing and strange case of whale population recovery," explains Jorge. "Most of the world's whale populations have been slow to recover from whaling. Almost all of them are under half the numbers there were before whaling. Except gray whales." The eastern, or Californian, population of gray whales (the ones that breed off Baja) is the only whale population that is increasing to a level thought to be close to their historic numbers. "It is faster than we can expect for this kind of animal with a long life cycle," Jorge explains. "In relatively few generations, it has recovered. And why this one and no others is another interesting question."

Attempts to answer interesting questions often expose hidden assumptions. That's the intriguing thing about Liz's genetic work. We assume that California gray whales have made a spectacular recovery because we've been told they were hunted to near-extinction by the late 1800s. Yet ironically the population estimates of how many whales there were in the first place come from the records of the hunting that decimated the population. In the case of gray whales, folks have looked at the number of barrels of whale oil sold and then calculated from that not only how many whales must have been killed but also the size of the original population. But whaling records are filled with uncertainties. Things like the number of whales that were struck and killed but not taken don't show up in the record of whale barrels, and there's always the possibility of falsification or lost records. Obviously, gray whales come in different sizes. Calves are far smaller than their parents, yet only one translation factor was used in going from barrels of oil to the number of whales.

Liz's work reveals a fundamentally different story. "Certainly some of these estimates that you see for the bottleneck population size just must be wrong," she says. "You'll see a hundred and twenty animals, a thousand animals. I think it was probably more like five thousand or seven thousand or ten thousand individuals left at the bottleneck. Because we can look at how many genetic types there are in the population today and say, OK, we can see that there are a given number of genetic types today. That means that there certainly must have been more than a hundred and twenty animals. It seems to me that all the evidence points to the fact that there were quite a large number of gray whales left, even at the bottleneck."

So what does it matter if a hundred years ago the gray whale population off North America hit a historic low of seven to ten thousand individuals rather than just a hundred and twenty? The short answer is, it matters plenty. The International Whaling Commission places a moratorium on commercial whaling until a population reaches 54 percent of its historic numbers. If, as Liz's genetic work is beginning to reveal, the prewhaling population of gray whales was three to five *times* larger than the estimates based on the whaling records, then gray whales are nowhere near to being recovered.

Liz's work bolsters a landmark 2003 study reported in the journal *Science* by her graduate supervisor, Steve Palumbi, professor of biological sciences at Hopkins Marine Station of Stanford University, and by Harvard University graduate student Joe Roman.[3] Using DNA analysis, they found that, before being decimated by nineteenth-century commercial whaling, humpback whales in the North Atlantic Ocean numbered about 250,000 individuals—more than ten times more than previous estimates based on historic whaling records. "It is well-known that hunting dramatically reduced all baleen whale populations, yet reliable estimates of former whale abundances are elusive," they write. "Whaling logbooks provide clues but may be incomplete, intentionally underreported, or fail to consider hunting loss." According to Steve, the worldwide population of humpbacks may have been as high as 1.5 million. "This is a real conundrum," he says. "Humpback whales, for example, were thought to have numbered about twenty thousand in the North Atlantic, and we're up to about ten thousand now, so at that rate, the IWC could allow countries to start killing humpbacks within the next decade. But if the historic population was really two hundred and forty thousand, as the genetics suggests, then we wouldn't be able to start whaling for another seventy to a hundred years."[4]

The size of historic whale populations isn't the only information scientists gain from studying genetics. Liz's research paints a new and captivating picture of the gray whales as highly versatile animals. Today we know them as bottom-feeders who leave giant ice-cream-scoop indentations on the ocean floor each time they filter a great gulp of mud through their baleen plates. "It's apparent that what we see today is a remnant of the whale population that once existed," Liz explains.

> The genetic work shows that population levels were pretty high, even back in the Pleistocene before the last glacial maximum, when the entire feeding range they used today would not have existed. So it's kind of remarkable. You know the Bering Strait was closed. The continental shelf was exposed. Yet they were thriving. So then you have to ask, What were they eating? Where were they going? And the answer may have been that they had different benthic feeding grounds. The answer may also have been that they were feeding in the water column, which they certainly can do. There were three healthy populations rather than the one that we have left. Their geographic range as well as their ecology was much broader than it is today.

This more diverse ecology may help to explain how some gray whales escaped the onslaught of whaling. "During the peak period of industrial whaling in the Baja lagoons, some whales may simply have chosen to go somewhere else," says Liz. "It may be that there was a large population that remained even at the very peak of whaling, but they chose either the Gulf of California or some other area in which to calve. That's one idea. The more I learn about gray whales, the more I realize that often the exceptions are the rule. So now we're seeing them breeding off the Channel Islands. We're seeing a lot of whales that are not necessarily traveling all the way down to Mexico, or that are traveling past the tip of the Baja Peninsula into the Gulf of California. It appears that they do have quite a lot of behavioral flexibility."

Discerning what is "normal" for gray whales isn't easy, but it's vital for predicting and for trying to ameliorate the effects of global environmental changes on remaining populations. Liz is in the process of analyzing three-hundred- to three-thousand-year-old ancient DNA samples from whale bones found in the archaeological middens—*really* old garbage piles—from an indigenous tribe in the Pacific Northwest. These data will provide a window into the genetic diversity present before whaling and can potentially be used to measure fine-scale population dynamics over time. She is also in search of ancient western Pacific and Atlantic gray whale samples. With the help of these historic data, she hopes to be able to model the past sizes not only of the California gray whale population but also those of the Western Pacific and the Atlantic gray whales. If, as her research suggests, those populations would have had to have been basically equivalent in size to account for the genetic diversity today, it would change our perception of gray whale conservation priorities. "We might not necessarily be as concerned that the eastern or California population had been decimated and hadn't recovered," she explains. "We would highlight the fact that we've entirely lost the Atlantic population, it's gone. And the Pacific population is an incredibly tiny fraction of what it once was."

The problem with the pervasive belief that California gray whales are a conservation success story is that it lulls us into complacency. If the population is recovered, there's nothing to worry about. We can stop whaling, and things will return to normal. Liz and her colleagues don't buy that story, particularly for gray whales, which in their life history span an incredible range of environments—from lagoons in Mexico to the Arctic environment, which is incredibly sensitive to climatic change. "What we're seeing now is this very, very rapid shift in many aspects of the Arctic ecosystem that are having direct impacts on this gray whale population," she explains. She points to a recent article in *Science* demonstrating that with climatic shifts in the Bering Sea, a lot of the food sources that were once available to gray whales have now shifted north.[5] There's only so far north they can go. And then that's it.

"All our focus has been on getting gray whales to recover from this insult from whaling," Liz says, "but there are these other really important stressors. Boat traffic is one, and noise pollution, the auditory threat, which cumulatively must have an enormous impact on gray whales." I think back to my litter-strewn drive to the Baja calving lagoons as Liz recounts her experiences of battling traffic in Southern California. "I go crazy from the traffic," she says, "and I'm not using my ears to migrate or communicate or listen to friends or mates who are far away." She worries about the massive concentrations of people living on the coasts and what it must be like to swim the gauntlet through the toxic runoff that drips from our streets into the sea. "I think the public eye is on how terrible whaling was and how these animals are recovering," she confides, "and less attention is on these chronic issues that are more deadly but perhaps less sexy. They kind of hold a mirror up to our own faces and all the effluent, if you want to call it that, of modern life."

An eye rises above the rail, jolting us back from the lull of a whale watching siesta. Kip flings himself against the side of the boat, stretching every muscle in his body in his attempts to make contact with the whale's head. He cries out in frustration. It is beyond his grasp. Our fellow traveler from Mexico City has more luck, his adult arm just enough longer to bring his hand down for a gray whale pat. We shriek and applaud and cheer as he returns to his seat, cold and soaking wet and grinning from ear to ear. We promise to e-mail photos. The whale swims off, and Fito slowly turns the boat back toward the shore. We settle in for a sun-kissed journey back to our launch site, sharing tales of the mother and calf, the pair of enthusiastic juveniles, the day of gray whale encounters already slipping into the realm of favorite family adventure stories.

Back on land, we skip rocks and hunt for shells on the beach while our guides load the boat back onto the trailer. I scramble over the dunes, happy to leave the pollution at the tide line behind and to marvel at the casual, remote feeling of the landscape. I feel the telltale crush of clamshells beneath my feet and look down to find that I am standing on a glittering white mountain of broken shells. I call the kids, and we walk inland, excitedly tracing the changes as the piles shift from one type and color of clamshell to another, filling our pockets with the shimmering shards. It feels like hunting for not-so-buried treasure.

I mention the piles to Enric Sala, a marine ecologist at the Center for

Marine Biodiversity and Conservation at the Scripps Institution of Oceanography, when we meet over muffins a few weeks later. "The fishermen have actually written the history of extinctions with those piles," he tells me, explaining how everything from turtles and sharks and rays and skates has been almost eliminated by fishing in the gray whale lagoons. For the past twenty years there's been an intense clam fishery.

> There are these piles of *almeja* and *catarina* clams. I don't remember the order exactly of extinction, but then suddenly the piles change to another color. That's a *pismo* clam. And you continue walking on top of these piles, and then there is a change in color again. So *pismos* disappear, and now you have the chocolate clam *(Megapitaria squalida),* and then the chocolates are almost gone, and then you have the mother of pearl. There have been these serial depletions in the lagoon. That's in the lagoon that the whale people believe is pristine. Because they only care about the whales; they only see the whales and they believe that the poor artisanal communities are in balance with the environment.

I am beginning to realize that one of the great challenges of conservation is that we can never see the whole picture. We pay attention to the things we recognize, the things that exist at physical and temporal scales that we can relate to as individuals. Whales live on a scale too big for us to comprehend. Tourism, even eco-tourism, places us in landscapes too foreign for us to read. I recognize the destruction that is litter but, as in the case of the clamshells, fail to see abundant evidence of overfishing even as I am hiking through it. "We are not used to thinking about the ocean in general. It is this out-there other world," agrees Boris Worm, marine biologist at Dalhousie University. "And we can't think of things in such a large scale. It is a big issue. Even we scientists tend to study things regionally."

I remember Hal Whitehead's ideas about time and scale. "The ocean operates on a lot of different scales, and in some ways, the longer ones are almost always the most important," he says. "There is no kind of prehuman stability. It was a very dynamic place, especially in the large scale, before we started messing around with it. But we have messed around with it in fundamental ways that override the natural variability. And that's pretty scary when you think about it."

Hal thinks that what we need to do is to stop setting benchmarks for what we'd like to conserve and start setting the benchmarks for our own behavior. Boris agrees. The author of several landmark studies on biodiversity loss in the world's oceans, Boris knows better than most the scale of global marine problems and feels compelled to sound the global alarm. Yet he also believes that we have the knowledge, moral responsibility, and collective capacity to solve them. Over coffee at a café near my home in Pacific Grove, California, Boris and I sit by a window with an expansive view of the Pacific Ocean. "You have to remember that with the great whales, we as humanity stopped whaling," he reminds me. "In fact, arguably, the first collective act of humanity was to save the great whales. That was the first thing we did as a global community. And it has actually worked, on a global scale. It hasn't worked for every species. Some of them are still threatened, like the Atlantic right whale or the blue whale, which seem not to be recovering. But other things, like the California gray whale, could be seen out our window right now." He excuses himself to answer a phone call, and I peer out the glass at the water beyond. It looks like unspoiled paradise. I hope to hell Boris is right.

14

The War on Fish

Ocean watching is a mysterious pastime. It's one of the few activities I can think of, other than the game of cricket, in which you can spend hours looking at something and still learn nothing about it. Swirl your fingertips across the ocean surface, and you can't tell if the water is fresh or foul. Is something in the ocean that shouldn't be there? Is something missing? Water, though transparent, hides its ills.

Marine ecologist Boris Worm knows more about the state of the world's oceans than most. His research focuses on the conservation of marine biodiversity on a global scale, and the causes and consequences of changes to these vast ecosystems. But even he finds it hard to "see." He illustrates this point by sharing the ritual of greeting the dawn. "My bedroom window back home in Nova Scotia overlooks the ocean," he tells me. "I wake up every morning, and I see this ocean view. The first thing I always feel is wonder." The ocean is such a vast place it makes him curious. It inspires him. Then his brain wakes up, and his sense of wonder is replaced by a profound feeling of ignorance. He thinks about just how little we know of the ocean. But he also appreciates the tremendous amount we have learned in just the last five to ten years. "It is just breathtaking to me, the new techniques we have and the global view we're taking," he says. "It is bringing about a very new vision of the oceans."

The trouble is, the more Boris learns, the more he worries. He is the co-author of the highly influential 2003 *Nature* study that found that 90 percent of the top predatory fish have vanished from the world's oceans.[1] In a

2006 issue of *Science,* he and an international group of ecologists and econ-
omists forecast the collapse of all currently fished species of seafood before
2048.[2] "Most of the technology we throw at fish is military technology," says
Daniel Pauly, director of the Fisheries Centre at the University of British Columbia. "All the acoustic equipment was developed, for example, in WWII
by the allies for chasing German submarines. The GPS, which is the global
positioning system, was developed during the Cold War to position things
and to study the Earth in great detail, and now this technology is available
to everybody to catch the last fish. . . . What I'm saying is that we're winning
it; we have won the war on fish."[3]

The large-scale industrialization of the fishing industry since World War II,
coupled with a global boom in oceanfront development and a rise in global
temperatures, is causing fish populations to plummet. "I now understand that
when I look out my window, I am looking at one of the most degraded, most
depleted, most unnatural shelf ecosystems in the whole world," says Boris.
"It is a sea without fish."

Wonder. Ignorance. Worry. This is how Boris begins each day.

My friend Jim is an avid birder. He caught the bug from his dad, one of the
first park naturalists in the California parks system. Earlier this year, Jim's
dad died, leaving behind a beloved collection of journals in which he'd metic-
ulously recorded the number and diversity of bird species he'd observed
throughout a five-decade-long professional career. What strikes Jim now, as
he sifts through his father's neatly penned entries, is how much has changed
just since his father's generation. It's not that Jim doesn't see enormous flocks
of birds. He does. It's just that what he considers "enormous" dwindles in
comparison to the astonishing numbers recorded in his father's steady hand.

It's a clear case of shifting baselines, what Daniel describes as the very human tendency to redefine what is "natural" according to one's own experi-

ence. The same phenomenon holds true, I can promise you, for the size of blue jeans over time. As long as you only keep a recent pair, you remain "normal" sized. Only those sadistic enough to save a pair from their college days are aware of how far a baseline can shift in a single lifetime.

Blue jeans are on my mind as I weave my way through the maze of low-slung wood-sided buildings that form the Scripps Institution of Oceanography in La Jolla, California. Jeans seem to be everywhere—lounging patiently on the backs of car seats while their surfers ride the waves; hugging tightly the bottoms of young graduate students or sagging gently behind older professors. I have come here to meet Enric Sala, who has been studying shifting baselines in the Gulf of California. He's wearing jeans. When I excuse myself to pop into the women's bathroom, I nearly crash into a tampon machine bedecked with a poster of a gorgeous, half-nude man provocatively fingering his denim fly. The jeans, the poster, the smell of the corridor: it's all powerfully reminiscent of my own student days in the early 1980s. It's as if the concept of "university" has remained unchanged in the past three decades. The same cannot be said for the state of the world's oceans.

"When I describe the underwater world, I am so passionate," Enric croons, his dark brown eyes shining with the pleasure of the thought. It is barely 7:00 a.m., and a day packed with meetings and impossible deadlines stretches far ahead of him, yet he patiently starts at the beginning. He starts with himself, a boy who fell in love with the sea. "I still remember the first time I read *The Silent World*," he says, describing the scene in Jacques Cousteau's epic book when Cousteau first puts his head underwater and leaves the noisy world of a crowded beach in France behind. He enters a new dimension that changes his life. "Unfortunately the sea is still the silent world," says Enric. "But now it is silent for a different reason. We have removed most of the large animals."

Enric is consumed with finding baselines. He wants to understand what the ocean was like before we started tinkering with it. Diving, he tells me, is his time machine. He dives often, traveling to some of the most remote places on the planet, hoping to discover the sea that time forgot. In 2005 he caught a glimpse. A thousand miles southwest of Hawaii on a remote string of islands in the Pacific he found paradise on Kingman Reef.

"Look at this coral forest," he exclaims, pointing to an image of the reef on his computer. The photo is so brilliant, so colorful, so bursting with life, it threatens to spill from the screen. "It's hundreds of years old. These table corals are up to four meters in diameter. This snapper is about two feet long. The visibility is two hundred feet. Sharks. Sharks everywhere. This is what a living coral reef looks like."

He pulls up another image of a reef and traces the outline of a coral head surrounded by algae with his finger. Four dazzling yellow-striped fish are nestled among the branches. It's breathtaking, and I'm about to say so when Enric scoffs: "This is what people think is beautiful. Coral reefs were called coral reefs because they had corals. Now they should be called seaweed reefs." His point is that we've become so used to seeing degraded coral reefs, we don't know what they're supposed to look like anymore. He clicks back to the glorious image of Kingman Reef and pauses for a moment to look. "It was a very emotional experience," he whispers, referring to his dive on a pristine reef. "I was well aware of the impacts of humans on the oceans, but not until I had that experience did I realize what we've lost."

Most of us have never seen a healthy living coral reef. The fact that we can't tell the "best" one from a mediocre one isn't surprising. The bigger concern is that fisheries managers, the people who are supposed to understand the ocean, have the same problem. "It's a classic problem for fisheries," says Enric. "Mexico, like many places, doesn't have detailed fisheries statistics, and people who are managing the fisheries are young, so they don't have a good baseline. They believe what is natural is what they first saw when they started their careers."

There are lots of clues, however, that fishing ain't what it used to be. Enric shows me an old black-and-white photo of a man sitting in a boat. He's completely surrounded by enormous fish called *Totoaba macdonaldi*, or "big croakers." The fish look close to six feet long, and Enric tells me they can exceed two hundred pounds each. They are so plentiful, the fisherman in the photos is harpooning them with a kitchen fork.

Totoaba were the first commercially important fish in the northern Gulf of California. Mexican fishers used to ship them overland from San Felipe to San Francisco and San Diego, marking the first time refrigerated cars were used. Today there are probably less than 300 individuals left. "The Mexican desert used to be full of dead carcasses, because in the beginning they used only the swim bladders," Enric explains. "And then they started catching the fish. And then of course all these dams were built in the United States and the water from the Colorado River was reduced to nothing. But this happened first. The removal of all these millions of fish."

Recently Enric attended a meeting with a well-respected fisheries scientist in Mexico, an old-timer who claimed that none of the fisheries in the Gulf of California is collapsing. "What about the totoaba?" Enric asked him. The fisheries scientist replied: "That's because of the lack of fresh water. When I started my career in 1968, they were already rare, so it cannot be fishing." Enric throws his head back and gestures so wildly his laptop jumps a foot closer to me. "What a ridiculous argument! This is the perfect example of shifting baselines. There are fisheries scientists like this."

Even marine biologists are susceptible. As Paul Dayton, an oceanographer at Scripps, puts it, "Each succeeding generation of biologists has markedly different expectations of what is natural because they study increasingly altered systems that bear less and less resemblance to the former, pre-exploitation versions."[4]

How little are you happy with? That's the question Enric poses whenever he manages to collar a fisheries manager or politician with influence on the marine environment. He tries to sway them by citing the example of Nassau groupers *(Epinephelus striatus),* large predatory fish that inhabit coral reefs throughout the tropical Atlantic. "Cuba is the country with the best fisheries statistics in the Caribbean," he explains. "Today they are catching a hundred tons per year of Nassau grouper. If this is all you know, you're going to be happy if for the next ten years you continue catching about a hundred tons. This is your baseline. If we go back twenty years ago, they were catching four hundred tons. So if you are a little older you won't be happy with a hundred tons. But if you go back to the sixties, it's fifteen hundred tons—fifteen times more than what you catch today. Are you happy with that?"

The Gulf of California is Mexico's most productive fishing zone. Yet fisheries statistics for the region are too unreliable to allow us to accurately chart the rate of its decline. Enric borrows techniques used by anthropologists to determine what's really happening to fish in the Gulf of California. He travels to small towns along Baja's vast coastline, collecting oral histories from fishers both young and old. "Here is what they were catching in the seventies," he explains, pointing to a brightly colored graph of steeply descending lines. "Goliath grouper, gulf grouper, sharks. This is what they are catching now. Trigger fish, small snapper, tile fish."

These fishers, like fishers all over the world, are "fishing down the food chain," as Daniel Pauly would say. They've switched from bigger fish to smaller fish, and as they've fished out the most desirable top predators, they've moved on to animals lower on the food chain. Humans are eating their way down the food chain, or the "trophic level," of the oceans, taking out the higher predators like white sea bass, halibut, cod, and striped bass first, then moving on to smaller midlevel predators and eventually down to invertebrates like jellyfish and plankton. Eventually, they will hit the bottom, leading to the collapse of fisheries in many places. You need look no farther than your local seafood restaurant to see this process in effect. "Americans wouldn't eat squid before," states Daniel.[5] "It was used as bait, and now Americans are eating bait. It has all kinds of fancy Asian names, but it's bait."

Enric knows all about seafood and restaurants. His parents owned a restaurant in his native Spain. But like most of us, they knew little about the real cost of fishing. "They probably spent twenty-five or thirty years selling shrimp and not knowing how shrimp was caught," he says. "People basically don't know because there is always fish in the market."

The more I speak with scientists about the magnitude of fishery collapse in the gulf and beyond, the tighter the knot grows in my stomach. Pure dread.

It's an ecological catastrophe. Yet everywhere I turn, restaurants, fast-food joints, and cooking shows are serving up sumptuous seafood delicacies. Take shrimp, the most commercially important fishery in the gulf. In Mexico it's called "pink gold" *(oro rosado)* because of its economic value in the international market.

Shrimp now outpaces canned tuna as America's favorite seafood. Last year more than a billion pounds of it were consumed in the United States, a record 4.2 pounds of shrimp for each of us. But what the couple enjoying a shrimp cocktail over a cozy table for two often doesn't realize is that shrimp is one of the most environmentally destructive things you can eat. For each pound eaten, somewhere between five and twelve pounds of other fish, shellfish, sea turtles, and even dolphins are caught and killed. This unintended bycatch is the result of heavy shrimp trawling nets being dragged along the ocean floor. Opponents of shrimping call these nets the world's most destructive fishing machines.

For the past two decades, the shrimp catch in Mexico has been dropping. Scientists blame climate change, overfishing, and a host of other environmental problems.[6] So why can we still buy cheap Mexican shrimp? "It's because the fishing lobby is very powerful," Enric tells me. "If you have a business and your business doesn't work, you know you're going to have to change your business. But what's happening with fisheries is that governments use taxpayer's money to subsidize an activity that is not profitable. So basically the fishermen are exploiting a resource that is not theirs. It is a common resource. And they are subsidized to overexploit it." When you shell out money to buy seafood, as he explains it, you're simply making the final payment on an item you've already paid for several times before.

"The shrimp fleet is in economic crisis," confirms Maria de los Angeles Carvajal, executive director of Conservation International's (CI) fifteen-year-old Gulf of California program. "The fleet owners realize they're overcapitalized, and many of them want to sell out anyway. Meanwhile, those who do remain in the business could double their income while harvesting sustainably."

For much of the past decade, CI has been searching for technological ways to reduce the destructive impact of shrimp trawlers. Because they have a mandate to conserve not only marine ecosystems but also the livelihoods of people who make their living off the sea, CI takes the tack that fishing isn't the problem. It's the *way* people fish that needs to change. "In the end we found a technology that can reduce 65 percent of bycatch that the shrimp trawlers are producing annually," Machangeles (as she prefers to be called) tells me when we meet in her office in Guaymas. Both fleet owners and Mexican authorities eventually approved the strategy, and in 2005 they announced a $64-million buyout of half the fleet. The government also launched a small pilot project in Sonora and Sinaloa, two states bordering the Sea of Cortez, to put the new gear on about twenty boats.

Finally, a fisheries issue with a promising ending. But not so fast. In 2006 the Mexican congress failed to approve the new budget to continue this program. "It was not a priority for the campaign in the June 2006 general election," Machangeles explains. "We needed 50 percent reduction of this fleet, and we have only 22 percent at this point. In a country like Mexico, and it is the same all over the world, fishing is not a significant contributor to the GNP." When it ceased to be a budget line item, the destruction of the gulf's marine biodiversity from shrimp trawling disappeared from the press. Like the bycatch it seeks to protect, the policy itself has become an unintended victim of a political process much bigger than itself.

Marine conservation is a messy business. It's intensely political. And personal. Machangeles remembers vividly the day in September 2002 when environmental groups, including CI, succeeded in getting the government to enforce an existing law to ban shrimp trawlers from the upper Gulf of California and the Colorado River Delta Biosphere Reserve. The reserve was created in June 1993 in large part to protect the habitat of totoaba and vaquita *(Phocoena sinus)*, one of the world's most endangered porpoises. "The most powerful people in this town are the owners of this fleet," Machangeles explains. "It was a big social crisis. I was worried about my family. My son and daughter are going to the schools with the sons and daughters of these people. Sometimes it is very difficult."

The gulf provides food and employment for nearly half of Mexico's burgeoning coastal population. Fishing and shrimp fleets here catch roughly half of all seafood consumed in Mexico. In the upper gulf, fishers protested the decision by blockading the highway leading from the reserve to the U.S. border, temporarily trapping Arizona tourists in the beach town of Puerto Peñasco, also known as Rocky Point. At that point, federal police ordered the protesters to disperse, and Mexican environment officials began negotiations to resolve the crisis. The agreement they reached allows 130 trawlers from local communities to resume fishing in the reserve under more restricted conditions.

The debate over how much, if any, fishing should be allowed in a marine reserve continues to rage in the gulf, as it does throughout the world. For Enric, the answer is clear. "Every fishery that is managed based on science, so far, has collapsed, or is going to collapse," he says. "And we are not buying any insurance. Marine reserves can provide this insurance." The advantage of no-take zones is that they allow fish to grow bigger. The bigger the fish, the greater the number of babies it will produce.

Allowing trawling to take place in the no-take area of the biosphere reserve in the upper gulf amounts to carrying too little insurance. The same is true for the large marine park off Loreto, in the southern gulf. Less than 1 percent of the park's sea surface area is closed to fishing. "Basically the no-take areas are too small to be seen on a map," Enric says. "Yet that park is one of the flagships for marine conservation in Mexico." Not surprisingly, the park hasn't been effective. "The fish population changes in the park are the same as the changes outside the park where there is no protection," Enric explains. "It's not doing what it's supposed to do in a time frame where no-take areas in other parts of the world have shown spectacular increases."

When I join Mike McGittigan of SeaWatch for breakfast and casual gossip with a group of real estate developers and luxury boat owners at a yacht club in La Paz, I see a successful entrepreneur and a charter boat operator who is comfortable in the world of money. But I also see Jacob Marley—a man tormented by hindsight. In the forty years that he's been diving in the Gulf of California, he's been given the ability to grasp the extraordinary richness of the gulf and to bear witness to its destruction. He has seen paradise lost.

"If I could have the insight and the knowledge that I have now, and be looking at those reefs that I was looking at in the early seventies," Mike says as he gazes beyond the table to the turquoise waters of the Bay of La Paz, "I would have documented things very differently. There used to be fish everywhere. Now it's all disappearing. I would have wanted to record it for history. I never imagined humans could do this to the aquarium of the world."

It's the speed of the destruction, and its global scope, that is so terrifying. "The absolutely frightening part is that I see no relief," says Mike. "The fishermen don't go away, even when there's nothing left. It takes major change to reverse the process. You've got to take the nets out of the water. You've got to stop fishing. You've got to truly protect areas."

Enric looks to Australia for inspiration. They have the largest network of marine-protected areas in the world, the Great Barrier Reef Marine Park, and in 2005 they closed 30 percent of that marine park to fishing. He also finds hope in Cabo Pulmo, a small national marine park near the tip of the Baja Peninsula. Just ten years after receiving protection, it's made a spectacular recovery. "Now it's one of the places where corals cover most of the bottom," he says. "And as soon as you go down you have these large gulf groupers who come to check you out. It is so wonderful. We are talking about fishes that can be between a hundred to a hundred and fifty pounds, literally as large as the windshield of a car. If you go down to the bottom and just stay there in the hole and try to breathe as slowly as possible, they come really, really close to you. These are very curious animals. It's such a wonderful place."

As Enric describes how such recovery occurs, I feel my chest begin to relax. I realize I've been holding my breath; too many interviews about ocean devastation have generated the anxious tension of a horror movie fest. These safe havens are so productive, he assures me, that even I would recognize the differences. Usually the animals inside the reserve grow larger first, and then you start seeing greater numbers. "You don't need to do any analysis or fish count," he says with a slow smile. "You go by diving, and in three years, you can see differences."

I close my eyes and try to imagine the underwater world as Enric has seen it—as close to pristine as is possible in today's world. How do we stop the war on fish, I wonder, if there are so few cease-fire zones? How can the ocean recover if we continue to load the places where fish feed and breed with high-powered fishing artillery? The questions sound hauntingly familiar, and suddenly I realize they're the same ones we're all asking about the war in Iraq. Instead of collateral damage, there's by-catch. In place of embedded reporters, there are marine biologists. And everywhere there are unspeakable numbers of dead and dying. Distanced from the problems, weary of the weight of such complicated, intractable dilemmas, I am numb. I turn away from head-lines about fisheries collapse just as I switch off the radio at the first report of a suicide bombing. Where does one find the resilience to carry on?

Resilience is something Boris Worm has thought a lot about. He believes the real economic costs of overfishing are profound. Overfishing makes the marine environment less resilient. It's not just that your favorite seafood is disappearing, it's that the whole ocean and, by extension, the whole world becomes unpredictable and unstable. "If you don't know whether your tourist resort is going to be viable next year because of a toxic algal bloom it brings about an uncertainty in life. It kind of shakes our fundamental assumptions about how the world should work," explains Boris. "I see people increasingly getting nervous about this. They are less likely to invest because they can't really infer from the past anything about the future."

A less resilient ocean and less predictable economic systems are twin ter-rors. Lack of resilience means that when disturbances happen, they not only happen more often, but they disrupt the ocean permanently. Things don't just bounce back. They stay changed. And they'll change again when the next disturbance happens. It becomes more and more an unknown and fright-ening path.

Boris dreams of getting the oceans back on track. Not back to what was, because he knows that doesn't work, but back to marine ecosystems that are robust enough to handle both natural disturbances and the kinds of insults we impose through overfishing, climate change, and all the rest of it. "The key term people use is *resilience*," he says, "which I kind of like because it brings up ideas of being a resilient person. You know, being strong but not static, to respond dynamically. Being able to recover once you've been sick or had an accident. All these things happen. We know that from our own lives."

I think of Boris greeting the ocean each morning with a full comprehension of its beauty and of the slow-motion disaster we have wreaked upon it. "The idea that humans are having a global impact on the planet as a whole is the biggest wake-up call we've had as a species," he tells me. "Maybe it's similar to what happened during World War II and the cold war. We were so struck that we founded the UN. They were kind of forced together by the realization that we have the ability to destroy ourselves. I think that is happening again now, in that we're realizing in a deep way that we are about to destroy the planet we live on. If you really realize what that means, it kind of changes you as a person."

15

Why Blue Whales Gotta Be Big

Every naturalist on a whale watching boat has a favorite way of explaining the gigantic proportions of blue whales. A blue whale's heart is as big as a Volkswagen. You could fit an elephant on its tongue. The body of a male blue whale is as long as three school buses and weighs as much as thirty thousand house cats. Eight kids could stretch across a tail. In the wonderful David Attenborough series, *Life of Mammals,* David wanders through the corridor of a blue whale skeleton while computer-animated graphics layer on fire hose–sized blood vessels and an aorta wide enough for a baby to crawl through. You get the picture. Blue whales are BIG. They are the biggest animals on earth. It's an awe-inspiring fact, but it's not news.

What is news, according to Don Croll, is *why* they're so damn big. Talking with Don, an ecologist, and his colleague Bernie Tershy, a research biologist in the Croll-Tershy lab at the University of California, Santa Cruz, feels a bit like chatting with an old married couple. Bernie makes the arrangements and gets me settled, while Don sits back in his chair and waits to pounce. With a squeak of a spring that is long overdue for a spritz of WD-40, he leans in, issues a proclamation, then tips back and waits. Without losing a beat, Bernie calmly poses a question, Don slightly tempers his assertion, and the two carry on in a rich exploratory conversation, finishing each other's sentences and building on each other's ideas.

Don and Bernie have been building on each other's ideas for twenty-three years, ever since they met while working on fin whales in the Gulf of California. Like most whale researchers, they've spent a lot of time watching whales at the surface. But not anymore. "Once we got an echo sounder," Don explains, "it was like suddenly the lights went on. All the action from the whale's perspective is going on down there."

An echo sounder is an orange shoebox-size device that is hull mounted to the bottom of a research boat. It uses sonar technology to send down "pings" of sound every few seconds. What Don and Bernie see as they scan a computer screen down in the cabin of a research boat in Loreto Bay in the Gulf of California is the sound waves reflecting off objects such as rocks, plants, or fish below. What may surprise you is that Bernie and Don aren't using the echo sounder to watch whales. They're watching krill.

Krill are a whale photographer's nightmare and a blue whale's dream. These tiny shrimplike crustaceans can form clouds so thick, it's impossible for a diver to see his hand, let alone shoot an underwater photo of a whale. Blue whales eat krill. In fact, that's all they eat. During peak feeding, a single individual may swallow four to six tons of the stuff each day. For many years researchers such as Christopher Clark have been puzzled by the abrupt way blue whales will suddenly stop feeding in the midst of what looks like a rich krill patch: "You'd be there in a Zodiac or research ship, and there'd be blue whales all around you," explains Chris. "And then they'd just disappear."

Why and where do blue whales go? As in most tales of intrigue the answers are complex, but Don and Bernie have a strong idea about where to look for clues. If you want to understand what blue whales are doing and why they're built so big, you've got to focus on the krill.

"Being large buys you two things," Don tells me. "It buys you the ability to store food, and it buys you the ability to process a lot of food quickly." If you're a runner, you've probably heard of the term *specific metabolic rate* (SMR). It refers to the energy consumed per unit of bodyweight. Runners care about it because they are trying to keep their bodies as light and as strong as possible. What's intriguing from a blue-whale-size perspective is that as a body gets larger, its mass specific metabolic rate only increases at 0.75. As a result, as an animal gets bigger, it really increases its capacity to store energy.

Blue whales can fast for longer periods and burn fewer calories than smaller whales because they are so big. That's critical because blue whales swim vast distances in pursuit of krill. Those studied in the Gulf of California depend on large patches of krill that can be found as far north as Alaska. When a blue whale finds a dense aggregate of krill, its enormous body size further enables it to eat a lot of it quickly.

"The bigger you are, the longer you can go without eating and the more easily you can travel around looking for food," adds Bernie. "If you're little like a shrew, you need to be a carnivore, because your metabolic rate is too high to get enough calories from vegetables." Bernie points me to a pivotal research paper by Wally Jarman on antelope in Africa.[1] "It's the best study that got us thinking about whales and size." Jarman demonstrated that as animals get bigger they can eat lower-quality food. Wildebeest and big kudus and other large antelope can eat low-quality grass, but they need to eat a lot of it. So they look for really big concentrations of food. They can afford to walk around these huge ranges looking for it because they're so big.

It's just the opposite for the dik-dik. These house cat–size antelope can't survive on grass because they aren't big enough. Their metabolic rate is too high. They can't afford to roam around looking for food, so they keep small territories instead. They guard those territories, keeping out other animals, so they can just nibble away on high-quality, nutrient-rich food. They stay more in the forest where there are always leaves on trees because they really need a constant supply. "Blue whales," Bernie explains, "are able to be more like wildebeest."

As I ponder the helpful parallel between wildebeest and blue whales, Don lets loose with a proclamation. "A blue whale opening its mouth to take in a school of krill is the biggest biomechanical event to happen on the planet." Immediately, I lose interest completely in plant-eating antelopes. "You've got a whale that's pushing a hundred feet that's accelerating at fifteen knots near a canyon that goes straight up. The total calories it takes for a whale to do a lunge are huge!"

Blue whales burn so many calories when they lunge-feed that they literally can't afford to feed in places where the density of krill is low. As soon as a patch starts to thin out, they head off in search of another cloud of "insanely thick" krill, as one of Don's graduate students calls it. She leads me to a computer and points to a deeply shaded section of the screen. "This krill patch formed a wall of red extending hundreds of meters in depth," she explains. "It was this thick from seventy-five meters to the bottom. Solid red!"

Don invites me to spend some time in the mouth of the blue whale skeleton outside his lab. He tells me he likes to stand there imagining the whale doing a lunge. The jaw dropping 90 degrees to the rostrum, and the big old jaws bowing out and disarticulating. Later, when we take a short break, I head outside and try it. Standing in the fog, in the mouth of a whale, I try to put an image to Don's words, and to those of his colleague, Alejandro Acevedo-Gutiérrez. A few weeks earlier Alejandro told me, "Some whales, like the bowhead whales *(Balaena mysticetus)* and the right whales, tend to feed by slowly swimming and what's called skimming—with their mouths open and filtering prey. It's by no means as active as what blue whales do, which is to launch to trap their prey. Imagine yourself as a human on a long field. Hold your breath and slowly walk and see how far you can get. That's a bowhead. Now, hold your breath, walk slowly, and then *sprint* for a few seconds and then go back to slow walking and then *sprint* and then. . . . That's a blue whale."

I realize as I stand in this cave of bones that blue whales are doing all this feeding underwater, sometimes going 740 feet deep and executing three, four, five lunges as they rocket toward the surface. They accomplish the biggest biomechanical event to happen on the planet. While holding their breath.

⁓

"If you had held a gun to Bernie's head ten years ago and told me to go and find some krill or you'd shoot, I probably would've spent some time sweating," Don confides. "But now, I could find it. I don't even think Bernie would sweat." Clearly Don is a confident krill spotter. But he's making a high-risk wager. If you want to find blue whales, you've got to find "insanely thick" krill. To find insanely thick krill, however, you have to understand the interconnections between wind-driven upwellings and tidal basins and a series of other oceanographic processes. Personally, I have trouble even with IKEA instructions, so when Don begins his explanation with "I would start with plate tectonics," I start to cringe.

But Don is undeterred. He and Bernie have pioneered a "wind to whale" theory that predicts the locations of the most highly productive places in the ocean. The motherlode of insanely rich krill. Mexico's Gulf of California is one of them. "Why are all these whales and seabirds there when they are?" asks Don. "How do blue whales come up with a migratory pattern that includes the Gulf of California? The whales are not what's unique about the gulf. What is unique about the gulf is its year-round productivity. That's what drives all that follows."

Those who like to listen to *Car Talk* on National Public Radio for the human-interest stories rather than for the mechanical advice may wish to skip the next few paragraphs. But the more I listen to Bernie and Don, the more spellbound I become by how various oceanographic processes combine with the physical geological history of the gulf to create the "happy series of coincidences," as Bernie puts it, that make the Gulf of California one of the richest most biodiverse marine environments on earth.

At first I feel like I'm eavesdropping on two gods, arguing over the finer points of how to make a sea. "The basics of it are that there's upwelling when the wind blows from the north on the one coast, and then there's upwelling when the wind blows from the south on the other coast, and then there are the really high tides," says one voice. "On top of that," booms the other, "because of the continental drift, you've got deep water that's located relatively close to shore. You've got accessibility of nutrient rich water that can be upwelled."

Upwelling is a basic process of the ocean, just as basic as the beating of your heart. The ocean is warmer at the surface than it is deep down. When winds push the surface water of the southern flowing California Current westward, cooler, nutrient-rich water from the depths rises to take its place.

Yet it is the gulf's extreme tides that boost it into superproductivity. "There is tidally driven upwelling in many places throughout the gulf," Don explains. "That means you get nutrient-rich upwelled water throughout the year, rather than seasonally the way we have it in Monterey. And that's one of the main reasons the gulf is green and rich in phytoplankton and productivity throughout the year."

I like trying to understand whales by looking at the processes of the ocean. I remember this feeling of having been given secret spy lenses through which to see things in new ways when I would go hiking in the Canadian Rockies with my geologist friend John. I liked the physical exertion as we climbed, the cold catch to my breath, the beauty of the rough and jumbled scree slopes of loose rock, the sound of the wind in the aspens. But John experienced the mountains in a whole other dimension. He'd point out the layers of sedimentary rock laid down as sands and muds and clay millions of years in the past. He'd look for the fossilized shells of clams that had died on an ancient inland sea and been folded high into a mountain ridge. He could appreciate the beauty of a diamond but was far more captivated by the extraordinary combination of pressure and heat that created it.

Don and Bernie are ecologists. Like John, they have a passion for systems. They're more interested in the oceanography—how the processes of tides and winds and upwellings and currents and canyons come together to make unique environments—than they are in what any one species does. From that perspective, a species is just responding to what is available. They see ocean systems where I see whales and fish.

It's the ability to read these processes that enables Don to find highly productive patches of krill, at presumably a high enough success rate for Bernie to be spared a bullet. "You have to think in two different spatial scales," he tells me. "Temporally, I know where I would look at any particular time of year. It's March, so I'd probably go down to the Gulf of California. But once I am down there, I know where I would look too. In Loreto, for example, I know what the bottom looks like. I know where it just drops off. So even at that small of a scale as the Bay of Loreto, there are places where I know we'll find krill."

This is impressive, but on some level it doesn't seem all that noteworthy. It strikes me the same way as the claim of, say, a mushroom collector who is able to find morels by knowing how to read the habitat and the seasons. The full significance of what Bernie and Don are doing sinks in later, when I am talking with Bruce Mate.

"You're using a land metaphor," he cautions. "On land, areas are stable. You can stake them out. They are very well delineated. But take a place like the Costa Rican Dome. The dome is not a subsurface structure; it's not a seamount. It's an upwelling *event* that varies in location from year to year."

I realize I have missed a critical notion. On one level krill occurs in predictable places, but on another level those *places* move. "The Costa Rican Dome is the confluence of three different currents coming together," explains Bruce. "So here you have a place that you can identify, but not it's exact latitude and longitude. It actually changes a little bit from year to year, depending on how the currents are going."

It's a thrilling and sobering thought. Ocean thinking requires more flexibility, different notions, than my experience allows. The ocean is dynamic. Krill is ephemeral.

16

What You Can See by Listening

How do whales contend with an environment where things are constantly moving? A world where nothing is fixed? I ponder this dilemma as I race around the house in search of an overdue library book. Story time starts in seven minutes. I want to take the bikes, not the car. And unless I return the missing book, we can't take out the armloads of new books I know Kip and Esmé will be eagerly selecting from the shelves. Esmé likes to shift things around. She's a collector. Well, more of a bagger, really. She loves to fill things with a potpourri of items. It is not unusual these days to lie back in bed and discover that my pillowcase has been stuffed with a carrot, a ring, a broken hotwheels race car, and a bread tag. I check all the likely locations—under the beds, in the cupboards, on top of the piano, in the armchair cushions—but no book. I do not think to look where Kip eventually finds it. Wrapped up in a blanket, resting, in Esmé's dolly bed.

Whales live in a vast dynamic environment; one that renders my experience of shifting objects nonexistent in both scale and complexity. Yet whales don't mill around like me, hoping they'll find something. "We can definitely say the whales aren't doing "random walks," Bruce Mate tells me. Bruce focuses his satellite tagging research exclusively on endangered large whales, which sadly, include blue whales.

Like acrobats meeting in the center of a tightrope, Bruce's satellite work and Don and Bernie's ecosystem analyses are coming from two different directions and reaching a common conclusion. Bernie and Don study highly productive krill patches to understand why blue whales show up where and when they do. Bruce tags and tracks individual blue whales to figure out their migratory patterns and determine what's so special, ecologically speaking, about the areas they frequent.

What Bruce first showed in right whales and then blue whales in the Pacific system is that if the density of prey changes, whales don't just mosey around in search of food. They'll work a patch and then just take off at some ungodly speed and go, say, twenty miles and find another patch. If they don't like that patch, they go to the next area. Or they might cycle back and come back to the first. He recently found the same dynamic among blue whales in Chile.

"We know what blue whales do in 'normal' years," Bruce explains. "We'll see whales in one of their typical year-to-year hotspots and then, when the wind dies down, two or three days go by, they're off. They don't just meander around slowly, back and forth, zigzagging along the coast. They go in a straight line for another place that's a traditional hotspot. And they move fast. They just aren't willing to fool around in low-density prey patches."

What does a hotspot look like to a whale, I wonder? I don't see the natural divisions of habitat in the ocean the way I've learned to identify them on land. I know what a meadow looks like. I can spot a pond. But it's difficult to wrap my mind around the ocean as a series of habitats when all I see is an endless expanse of blue water. I ask Bruce if he can see hotspots. If he can predict the route a blue whale will take from kelp patch to kelp patch. "Definitely not," he answers. "There is no universal consensus. You can have three blue whales in the same feeding place, and then when things settle down there, say they've cleaned it out, or the weather pattern changes so its not productive for them, one will go north. One will go south. And none stays in the same place."

Bruce thinks that large baleen whales have a repertoire of familiar options that vary from individual to individual. It puts me in mind of my favorite neighborhood restaurants: Passionfish, the Red House, Gianni's Pizza. Although the scale of the analogy is way off, I play with the idea that each is a highly productive patch of food and that whales, like me, know the ones that are reliably open on certain days or at particular times. I may visit all of them over the course of a month, but not necessarily in any particular order.

"Their mothers have lugged them around for a year and shown them the turf," Bruce replies when I ask him how baby blue whales find the food patches. "What they start out knowing is what mom knows, and maybe not all what she knows, because they may not have to explore the whole range of her options, if it's a good year."

A good year? What about a bad year? Remember the *annus horribilis* Queen Elizabeth described in 1992? Marital breakdowns for Charles, Anne, and Andrew. A fire at Windsor castle. It's hard enough to relate when a monarch is having a rough time, but what constitutes a bad year for a whale?

Bruce has met a lot of people who tell him they'd like to be reincarnated as a whale. He doesn't share that sentiment. "It's not a lay-back, fun-in-the-sun, get-a-tan kind of lifestyle," he says. "It's hard work. In an El Niño year, half the blue whales we saw were emaciated. We could see their ribs and vertebrae. Part of the advantage of being big is that you can live through some pretty trying circumstances. But you don't live through it healthily enough or with enough energy reserves to produce calves that year, or to feed your calf, or even to conceive a calf for the following year."

The beauty of Bruce's satellite tagging work is that it provides a deeper understanding of what individual whales do at specific times of the year but also of what's happening to whole populations over multiple years. Blue whales, it turns out, are highly focused eating machines that travel in marathon beelines between rich patches of food that are both likely to be there and impossible to guarantee. These whales can't afford to eat just any krill; it has to be a patch of krill sufficiently dense to cover the calories it cost to both find it and consume it in gigantic lunging mouthfuls. And they don't just live anywhere. And the ocean isn't the same everywhere. Different populations of blue whales have their own enormous but distinct ranges.

How the hell do they live like this? (I'm starting to sound like Don.) I really get that it's all about the krill. I appreciate the elegance of how Bernie and Don weave data about upwellings and tides and winds and undersea geomorphology and other ocean processes to predict the location of krill patches. But how do the *whales* do it? How do they find patches of thick krill in an ocean environment that is fluid and ever-changing in a way we never experience on land?

Doc White once found a line of krill twenty miles long and sixty-five yards wide off Mag Bay on the west coast of Baja. "We opened up a pair of Ceci's pantyhose with a coat hanger and used it as a krill net," he says, leaving me to ponder why his wife would bother to bring such things to sea. They strained out tiny krill and tuna crabs by the fistful. A few moments later they heard a crunch crunch crunch sound and turned to see a minke whale *(Balaenoptera acutorostrata)* feasting its way through the center of the soup. "How in the hell did that whale find the krill?" he asks. "How do any of these whales do it?

I bounce Bruce's idea, that blue whale calves learn krill locations from their mothers, off Doc and pause for dramatic effect. "But that's not what everybody thinks," I tease, momentarily smug at the prospect of pitting one internationally respected whale researcher's ideas against another's. "Christopher Clark says the patches are too ephemeral; the whales can't count on them being in the same locations." "So how does he think blue whales find krill patches?" Doc asks. "By listening," I say. "Blue whales find krill by listening."

It's such an intriguing and complex idea, I feel a little shy even trying to explain it. It's kind of like trying to paraphrase Stephen Hawking's *A Brief History of Time.* But its originator, Christopher Clark, is such an inspiring communicator, I'll give it a go. According to Chris, blue whales, fin whales, and humpback whales hear differences in the structure of the three-dimensional ambient noise of the ocean. "Imagine that you are a blue whale swimming northwest of Bermuda by a hundred miles," he suggests. "You'd be able to hear the difference between something to your north and something to your south. To the north, you would actually be able to hear the rush and motion and the turbulence caused by the Gulf Stream. And to the south, toward Bermuda, you'd be hearing the reflection and the illumination of the ambient noise of that mass, of that structure of the island."

I mull over what it might be like to hear ocean currents and reflections off islands when Chris suddenly kicks the guided-vision exercise up a notch.

> Now imagine being half a mile down and spinning very slowly and focusing your attention in a 360-degree circle. You might move your head back and forth in order to tune your ears. If you are listening, especially in the low-frequency ranges that blue whales are so good at, you would hear the noise from this river of warm water that's moving through the colder water of the Atlantic Ocean. And all the internal waves, because it's basically a very slow-moving large stream, you'd be able to hear that too. The sound is being radiated by the stream and is then reflecting off the underwater surfaces in the ocean.

People working for the U.S. Navy's antisubmarine listening system would be far more successful than I am with these scenarios. It's their job to figure out what makes noise and how sound propagates in the ocean, to distinguish the sound of an enemy submarine from an underwater volcano. With the end of the Cold War, Chris gained access to this system and combined it with customized systems developed at Cornell University. With this high-tech equipment he can "see" great whales communicating over *thousands* of square miles of ocean by using sound. He can chart their movement as acoustic herds. "It's like using the Hubble telescope of ocean acoustics right now," says Chris. "The navy had heard the whales for years but hadn't focused on the significance of their low voices."

Meeting people who have spent twenty-five or thirty years as acoustic analysts in these stations made a tremendous impression on Chris. "They're the closest I have ever come to people who are day in and day out spending their time either listening to or analyzing or being immersed in the noises of the ocean," he explains. "These people would say, if you put a good pair of headphones on me, and I have a good pair of hydrophones to listen to, I can listen to the ocean and I'll tell you exactly where I am." Chris didn't believe them. It was too amazing. But it was true. "This person isn't making any sound—this is a human—but they can do it simply by listening." This is the closest a human can come to sensing the ocean like a whale.

Chris tells me that when he is listening to the ocean, he is constantly creating an acoustic scene. He imagines how the hard structure of the rocks and the soft tissue of the ocean—the water—work together. "It's this dependence on the acoustic environment that the whales are married to," he says. "This is their fate. This is the way life is. And the acoustic part of their brains is ten times proportionately the size of ours."

Chris thinks that very thick krill patches may create subtle differences in the ambient noise of the ocean that blue whales can hear. It's also possible that blue whales listen for something completely different. Something ferocious. It's an idea that came to Chris when he was cruising through the research stacks and talking with people who know about ocean acoustics. It is possible that blue whales listen not for the sounds that animals make in the ocean but for the sounds of the ocean itself. Papers have been written about the seasonal coming and going of ambient noise levels in the ocean, and Chris thinks that blue whales may hear these. He thinks the whales are listening for ocean storms.

In the springtime, because of the rotation of the earth and the warming of different masses of the atmosphere and the water, massive storms are generated at relatively high latitudes. The storms pump waves that then propagate from high latitudes down into midlatitudes and the low latitudes.

Surfers track seasonal storms, often by using the National Oceanic and Atmospheric Association (NOAA) buoy website. They'll log onto the website, figure out when there will be a wave system, and decide to go surfing in three days.

The waves actually create a change in the low-frequency ambient noise signature of the ocean. They announce the advent of the storms of springtime. The storms are associated with the offshore winds that blow off the top layer of water. The end result? Upwelling. What Chris is telling me, I realize, is that blue whales might actually be able to *hear* what Bernie and Don are *looking* for—ocean processes that drive upwellings that result in those insanely thick krill patches.

There's one more way that blue whales might find krill, according to Chris. Their friends might tell them where it is. I'm relieved to be back in a realm of communication to which I can relate. My friends are champs at telling me when they've found good places to eat too. But why would blue whales do this, especially given how challenging it is to find suitable densities of krill? Wouldn't it be smarter to find a patch, tuck in, and keep mum about it?

"I think they use their very low-frequency voices to announce places where there is food," Chris says. He's observed that the distribution of singers is closely matched to the distribution of krill. "It would obviously be advantageous for a male to announce that he's found a patch of food to get access to a female. But it's also very realistic to think that the food is so sparsely distributed that it ultimately benefits you to announce the presence of a food patch to others, because if everybody does this, then everybody's assured of getting a minimal amount of food, and you might get more than that."

I grew up with the image of blue whales as mostly solitary animals. That's still the way they are characterized on the Web and in popular whale watching books. I don't even think there is a word to describe a large congregation of blue whales—a "flood," perhaps? Chris paints a very different picture of blue whales. They are social beasts. "If you went along the grand banks off Newfoundland and Nova Scotia in the late summer, the place would just be a lineup of blue whales," he explains. "That's where their food is. That's where they're singing. It's like a frog pond up there. You literally can't hear yourself think when all the blue whales are singing."

The ironic thing about blue whale society is that it occurs at a scale too grand for us to see. The distances between the whales are so vast that the connections between individuals become invisible. The "frog pond" of blue whales Chris describes to me belongs to a community that stretches all the way from the submarine canyon off the south of Nantucket all the way around Nova Scotia and up toward the Labrador Sea. "Does it make any difference that one whale is singing off Bermuda and one is singing off the Grand Banks?" Chris reflects. "Do they pay attention to one another? I don't know. But they certainly seem to coordinate a lot of their behaviors. Over a period of about a week, they'll all just essentially move out of an area and go to another. Now, whether it's because they've run out of food and one by one they all decide

to head northeast or something, I don't know, but I think it's a social unit."
Chris says he and other researchers have been thinking about whale herds at
the wrong scales. They've been constrained by the limitations of their
recording and visual observation abilities. "Suddenly you realize that the
whale's sense of scale is ocean basin sized—one song note is twenty miles
long and can illuminate the entire basin," he says. "It's an acoustic herd on
a scale that is just enormous."

It is an irony that just as researchers are gaining new ways of understanding
the links between whales and oceanographic features by listening to their songs,
what is being heard is a rising tide of "ocean smog"—the collective noises
from shipping traffic, oil and gas exploration and production, and recreational
traffic. And every decade the amount of noise is doubling.[1] "Whales have
very traditional feeding grounds, and their migratory routes along coastlines
have become incredibly noisy, urbanized habitats," Chris explains. "Acoustic
smog is shrinking their world." Chris conjures up an image of a world rapidly enveloped in fog.

Suddenly, in less than a century, there are all these unnatural storms showing
up. They don't follow any rhyme or reason like the storms of the spring would.
There are areas of your world where suddenly it's so noisy that basically you're
in a cloud. Year after year, one by one, all the coastlines of the northern hemisphere have been basically immersed in and clouded in this fog. It's not constant day and night, but it certainly is persistent. It's equivalent to that scene
you get from the satellite or the space station, looking down and seeing all the
lights from the earth. That is what we're seeing in the water. That visualiza-
tion is a direct analog for what's in the water. I could paint thick lines of red
going from all the major cities and ports of the world going back and forth
across the ocean. And suddenly your acoustic habitat, your acoustic ecology,
is just being literally eroded.

I ask Chris if the acoustic analysts he knows have experienced the impact
of undersea noise pollution. "Oh, yeah!" he answers. "They get acoustic white-
outs." He tells me the mouth of the Gibraltar is so loud, it's like putting your-

self in the middle of an expressway. "There are places and times when you are listening and a fishing fleet would come out and you would be just surrounded by a dozen boats," Chris explains. "They may be fifty or a hundred miles apart. But the collective mass of those boats spread out over hundreds of miles is like turning bright lights on. Suddenly it's like it's all bleached. You just can't hear anything."

But the worst stuff is the seismic activity that accompanies oil and gas exploration. Pull out a world map and look at the Denmark Strait between Iceland and eastern Greenland, or the Norwegian Sea where the Brits and the Norwegians are engaged in oil and gas development.

> You put yourself in that environment, where you have a seismic profiler exploding every twelve seconds. It's like being in the Grand Canyon when someone sets off a hand grenade. I've watched these things for days on end, and they will literally obliterate the acoustic environment. In fact, the navy people can't operate. And you think, God, these animals are there. This is their sole mechanism for surviving—finding food and detecting predators and finding each other. This is happening during their summer feeding period. And it's not just happening for one or two days; it's happening for months.

Not a single species of marine mammal or fish is naturally deaf. Sound is of critical importance to them. Chris and his colleagues have statistics for both the West and East Coasts of the United States showing that the levels of noise are now chronic enough and persistent enough that they would not meet the Occupational Safety and Health Administration (OSHA) standards required to prevent hearing loss in human beings. "You would have to put headphones on all these fish and seals and whales to protect their ears," he says with a wan smile. "We are now getting to the point where we may be doing permanent damage to entire populations of animals and not knowing it. We are raising the tide of noise to levels they can't compensate for."

In the past decade, I have met a number of scientists, like Chris, who have an extraordinary talent for seeing the natural world in ways the rest of us can't. It is a gift I envy. But I know it is a double-edged sword. What must it be like to glimpse the full wonder of the acoustic tapestry that is the world's oceans and to be one of only a few who understand its glory? To be almost alone in comprehending the horror of its destruction and compelled to confront a problem whose enormity appears overwhelming. Sometimes, when I am talking with scientists who study endangered species or threatened habitats, I feel as if I'm conversing with oncologists in the midst of a cancer epidemic, individuals whose job it is to record the exponential loss of something they hold dear.

Jane Goodall wrote about this in her book *Reason for Hope.* She describes the solace she finds in her religious and spiritual beliefs and in her knowledge, as a scientist, of the capacity of primate societies to evolve in new directions. I think Chris finds hope in his love of the natural world. "It's the whole ocean that is beautiful," he tells me.

It's not just the humpback whale that's on the postcard jumping out of the water. That may be magnificent. But to me, it's the wholeness of it that is so spectacular. And inspiring. We have some almost iconic species, like humpback whales, which for whatever reason we spend a disproportionate amount of time paying attention to. It's the whole we need to be considering. Because we aren't smart enough to know how to solve it one little thing at a time. If we go around picking out each of our little poster children, we're going to miss the big picture. And if we miss the big picture, we've got it wrong.

17

What You Can Learn from the Dead

Sometimes when I'm trapped in a long lineup, as I am today in an attempt to register Kip for swimming lessons, I while away the time by mentally compiling a list of the people I know with "special" talents: the fellow who makes the greatest three-layer dip; the woman who can spontaneously recite an appropriate line of poetry, no matter the occasion; the kid who rides a unicycle in our local parade; the gal with the inside track on the best income tax deductions. I vary the order, but there is one person who is always near the top. He is Craig Smith, the guy who knows how to sink whales.

"You need to use a lot of weight," he advises me when I ask him for a quick cook's tour of the deed. "With one whale, the largest whale we sank, actually, we put a lot of ballast on the tail, and it wouldn't quite sink. But then we towed it. With the weight of the tail underwater, it acted kind of like a hydrofoil, and it pulled it down a few meters. It compressed the gas in the carcass, and it sank immediately. A thirty-five-ton gray whale. That was quite a challenge." Craig pauses a moment and then mutters an afterthought. "Shooting it to de-gas it doesn't really do anything. Ship crews love to blaze away at dead whales. At least American ships crews do. The European crews don't have guns."

These are the talents they don't tell you about when you tick "marine biologist" off on the high school career form. Craig is an oceanographer at the University of Hawaii at Manoa. He's also a "whale fall" expert. *Whale fall* is the term used for a 30- to 160-ton whale carcass that is sinking to the ocean floor. The majority of whales fall naturally to the depths by themselves. Relatively few wash ashore. And only a handful or so of them are collected from beaches by Craig and his colleagues and then sunk, with the express purpose of studying the life forms that colonize them. In so doing, they've discovered new species that only exist on the bodies of dead whales. In the vast expanse of the seafloor, whale corpses turn out to be vibrant oases of life.

"To be honest, when we started working on whale falls, we didn't expect to find a lot of animals that seem to require whale falls," says Craig. "We also didn't expect the communities on whale falls to be particularly diverse. We are finding that whale skeletons support one of the most diverse hard substrate assemblages in the deep sea, and from what we can tell a significant number of those species have the potential to be whale fall endemics. We can't say for sure that they only live on whale falls, but at this point I believe there are twenty-eight species that have not been collected anywhere else."

How on earth does a species become so specialized that it only exists on the body of a dead whale at the bottom of the sea? Such extraordinary intricacies of the natural world are so thrilling, so worthy of shouting from the rooftops, that they make me feel like bursting into a kindergarten classroom and leading the kids in an evolutionary camp song. "There's a whale at the bottom of the sea. There's a whale at the bottom of the sea, and she's covered by strange new species." (OK, the meter's a bit off, but you get the drift.)

The desire to glorify such beasts grows even stronger the moment you lay your eyes on one. Bone-eating zombie worms *(Osedax sp.)*, for instance, are unlike any other animal. They look like tiny flowers poking out of the skeleton of a dead whale. They have no eyes, no legs, no mouths, no stomachs. What they do have is colorful feathery plumes and green "roots." The plumes act as gills and are connected to a muscular trunk. Like a retreating sock puppet, the plumes are sucked back into a transparent tube if the worms are disturbed. At the other end of the trunk, hidden inside the bones of the dead whale, the body widens to form a large sac laden with eggs.

"We've been collecting bone-eating zombie worms for years," says Craig, "but we couldn't get anybody to identify them. They were too weird for people to recognize. Jim Barry at the Monterey Bay Aquarium Research Institute

(MBARI) collected some off the coast of Oregon some years ago as well. Then Bob Vrijenhoek at MBARI and his collaborators found some big ones on a deeper whale fall on Monterey Canyon, and then they were able to describe them in *Science* in 2004."[1]

The scientists were initially puzzled by the fact that all the worms they collected were females. While examining the two- to seven-centimeter-long female worms under a microscope, they discovered that most of the females had dozens of microscopic male worms living *inside* their bodies. These male worms look as if they haven't developed past their larval stage. Their bodies still contain bits of yolk, but they also contain copious quantities of sperm.

Bob likens these worms to dandelions—a weedy species that grows rapidly, makes lots of eggs, and disperses far and wide. This strategy makes sense, given that these worms can only live on dead whales. After a whale skeleton has been consumed, all the worms at that site will die off. Before this happens, they must release enough eggs or larvae so that even a tiny proportion will survive and be transported by the ocean currents to colonize another whale carcass.

To appreciate how unique bone-eating zombie worms are, though, it would be helpful to take a moment to pop outside and peer at the sun. Every animal we normally think of relies on the sun for energy. Through the complex process of photosynthesis, plants convert sunlight, water, and carbon dioxide into sugars, which are eaten by plant eaters who, in turn, are consumed by carnivores.

Bone-eating zombie worms defy the norms of the photosynthetic world. Inside their greenish "roots" dwell symbiotic bacteria capable of digesting the fat-rich marrow in whale bones and supplying the worm with this chemical form of nutrition.[2] Chemosynthesis, as this form of productivity is called, is one of the major biological discoveries of the century. The first clues of its existence were uncovered in 1977, when scientists using the deep-sea submersible *Alvin* discovered a haven of densely packed animals in the sunless depths around hydrothermal vents 8,500 feet beneath the surface of the Pacific Ocean. It was later realized that these large and numerous animals were obtaining their energy from chemical-rich fluids generated by volcanic processes on mid-ocean ridges, the fifty-thousand-mile undersea mountain chain that encircles the globe and marks the edges of earth's tectonic plates. Instead of photosynthetic plants, chemosynthetic microbes comprise the base of this food chain. Bacteria in the water around the vents use carbon dioxide in the presence of other dissolved gases, such as hydrogen sulfide, to fuel the manufacture of sugar. Even more remarkable, the bacteria were found not only in the water but also living symbiotically in the tissues of the numerous newly discovered worms, clams, and mussels found around the vents.

"One of the things that interested us about whale falls when we found the first one on the seafloor is that they had a variety of species that looked like the kinds of animals that occurred in hydrothermal vents," explains Craig. "Animals like giant white clams and mussels. As we've studied whale falls more, we find there is overlap. There are a number of species that are shared, for example, between whale falls off Southern California and hydrothermal vents fifteen hundred kilometers away. It is quite conceivable that these animals are using whale falls as stepping-stones between widely separated hydrothermal vent systems."

The term *stepping-stones* conjures an image of little girls with pigtails crossing a gently babbling brook. What Craig is describing couldn't be more different. If they are lucky, the larval animals that are carried by currents along the California gray whale migration route from one gray whale skeleton to another ten miles away will have offspring that eventually end up in a hydrothermal vent spewing mineral-rich water at temperatures as high as 800°F. This water doesn't boil despite the scorching heat. At an average depth of seven thousand feet, the incredible pressure from the tremendous weight of the ocean above significantly expands the thermal range at which water remains liquid. The fact that animals thrive in such extreme environments shatters all previous conceptions of where life on earth can exist.

Craig believes that some species of mussels hitchhiked from shallow waters to great depths on sinking whales. "The mussels on whale falls are closely related to the giant mussels that were originally discovered at hydrothermal vents," he tells me. "In fact, they are all in the same subfamily. The ones on whales are quite small, but they use sulfide rather than sunlight as an energy source. They have chemoautotrophic bacteria in their tissues. Over evolutionary time it seems that whale falls may have provided stepping-stones for this subfamily of mussels that probably originated in shallow water to move down to the deep sea using whale falls and wood falls as stepping-stones and then radiating out to other sulfide-rich habitats on the deep seafloor."

It's a romantic notion. The ships lost by past explorers while crisscrossing the globe in search of new lands provided the wood falls that now nourish new communities of life in the deep ocean.

Whale falls, nevertheless, are the real jackpot. Food in the deep sea is extremely hard to come by. Until about 130 years ago, scholars believed that no life could exist in the deep ocean. The abyss was simply too dark and cold to sustain life. Today we know that nutrition drifts down from the sunlit surface zone of the ocean to the deep as fish and invertebrate fecal pellets and other aggregates that scientists call "marine snow." It would take up to two hundred years of marine snowfall to equal the food energy supplied by a single forty-ton gray whale carcass. Craig suspects that the Gulf of California has well-developed whale fall communities because of the abundance and diversity of whales species that frequent the region.

Just how long it takes for a whale to be consumed depends on its size and how calcified its bones are. As with any burial at sea, decomposition proceeds through predictable stages. First on the scene are mobile scavengers like hagfish *(Eptatretus deani)*, sleeper sharks *(Somniosus pacificus)*, and about forty more species of fish and invertebrates. They work fast, collectively removing more than 90 percent of the soft tissue usually within four months of a gray whale's arrival on the seafloor.

With the bones chewed clean, the path is open for the next stage of scavengers, about twenty different species of opportunistic polychaete worms, mollusks, and crustaceans that swarm the nutrient-rich bones and the seafloor sediment immediately surrounding the skeleton in dense assemblages of up to 40,000 individuals per square yard. This phase typically lasts up to two years.

The bone-eating zombie worms settle in during the third stage, the sulfophilic (or chemoautotrophic) stage in which the skeleton emits sulfide from the microbial anaerobic breakdown of lipids trapped within the bones. More than two hundred different invertebrate species have been identified on whales at this, the longest of the three stages.[3] Ten of these species also occur at hydrothermal vents, and twelve are found at "cold seeps," regions where methane and sulfide-rich fluids seep out from the ocean floor, supporting chemosynthetic ecosystems of tubeworms, mussels, shrimps, and other animals.[4] For big whales, like blue or fin whales, this stage can last at least seventy years. But for a smaller whale, say, a juvenile gray whale, it may last only three or four years.

So far Craig has identified 407 different species living on whale falls—an astonishing number, given that whale falls are difficult to find and are thus the least studied chemosynthetic habitats in the deep sea. This number rivals that of the global species known to inhabit hydrothermal vents, which are more intensively studied. And it represents substantially more species than identified at cold seeps.

Using DNA samples from animals collected from the deep, Craig and other benthic (deep-sea) ecologists piece together the fascinating evolutionary relationships existing between life forms in these chemosynthetic communities. "The molecular data are often useful for very oddball animals in telling you what they are most closely related to," explains Craig. It turns out that the most common ancestor to bone-eating zombie worms lived roughly 42 million years ago, about the same time whales themselves first evolved. Craig believes that giant tube worms at hydrothermal vents may have originated from shallow-water species using whale falls as a food source along the way. "It's conceivable that whale worms first appeared on whales and later adapted to deep-sea hydrothermal vents," he says.

It's tempting to think of the bizarre creatures in the deep ocean as existing in some kind of never-never land. The deep ocean is so alien, so impenetrable, that despite impressive advances in deep-sea research, the oft-quoted phrase "we know more about the dark side of the moon than we do about the deep ocean" probably still holds true. "The deep sea *is* remote and disconnected from our lives," agrees Craig. "As an oceanographer, I've spent many years at sea in the middle of the ocean. We go out for months, and you usually don't see other vessels. It is easy to believe that there is no human impact in the middle of the ocean."

Unfortunately, he knows this perception is false. "I think there is a misconception that the deep sea and all its resources are so vast, we can't possibly impact them through fishing activities," he says. "This is wrong." Bottom trawling is the practice of dragging heavy fishing nets along the seafloor. What the fishers are trawling for are deep-sea delicacies like orange roughy, a medium-size deep ocean fish with firm, tasty white flesh. The fishing method, however, is indiscriminate. It destroys most everything in its path, leading environmental organizations to label it the ocean equivalent of clear-cutting forests. "What I think most people don't realize," Craig explains, "is that in most of the world's oceans, virtually every square meter that can be trawled on the slope and the shelf without losing fishing gear has already been trawled. It is quite a remarkable realization."

Craig's conservation efforts focus on the vast and poorly understood deep sea, where high diversity, fragile habitats, and slow recovery rates render human activities, such as trawling, especially damaging. These days he's in a most unusual race. Craig is trying to better understand the distribution of animals living in the deep sea, the largest ecosystem on earth, with the goal of designing marine protected areas (MPAs) that the international seabed authority can then set up. The deep sea is such a tremendous reservoir of biodiversity and evolutionary novelty that 90 percent of the species collected in a typical mud sample from the abyss are new to science. Yet even as Craig makes a new identification, he is aware that that species may be in the throes of becoming extinct. It's a question of time lag. "If you do some very simple but robust conservation modeling to look at the impact of whaling," he says, "you come to the conclusion that in parts of the ocean, like the North Atlantic, the vast reduction in the abundance of whales that occurred in the mid- to late 1800s through whaling probably has caused or is causing species extinctions at the deep seafloor thousands of meters below in a whale fall ecosystem nobody dreamed existed until very recently."

Bone-eating zombie worms feasting on a whale skeleton that takes one hundred years to decompose, for instance, may have sailed through the peak periods of whaling, only to experience problems now, when they find that the whale skeletons that should have accumulated are missing. Today many populations of whales are so sparse that the distances between the stepping-stones they will provide when they die may be too widely spaced for larval animals to cross.

"If you want to preserve the diversity of whale fall communities, then you need marine protected areas to protect whales," says Craig. But it's not as simple as just figuring out where whales travel at the surface. Much like a Rubik's cube, Craig's plans for marine protected areas embody a three-dimensional vision of the ocean. "One of the ramifications of a connection between living whales at the surface and the deep seafloor is that you need to design marine protected areas that go all the way from the surface to the deep seafloor," he explains. "You may want to prohibit deep-bottom trawling in an area where you have whale populations protected at the surface ocean. That's required to maintain the whale fall ecosystems."

Like all good planners, Craig is also focused on the next challenge. In the deep sea, that is mining. Deep-sea nodules are potential sources of manganese, nickel, copper, and, most important, cobalt. Cobalt is considered a "strategic" metal. It is used, for example, to make alloys for jet engines. Countries are staking mining claims in the middle of ocean in part to protect their strategic interests—they don't want to be dependent on other countries for important metals. There appears to be an element of national pride involved. "It's kind of like the space program," says Craig. "It doesn't make the United States any money, but it has provided a lot of national pride. I'm convinced that if some of the countries that are doing deep-sea exploration could do it at no loss, or even at a moderate loss, they would take up mining so they could say they're the first country in the world to do it." Craig believes the technology to begin mining the deep sea will be in place within the next ten to twenty years. "It will be very devastating," he explains. "It will strip-mine tens of thousands of square kilometers of seafloor and cause a substantial disturbance in the biota [flora and fauna]. We need to get a better understanding of the distribution of the animals living at deep abyssal seafloors so we can safeguard against extinctions from large-scale mining if and when it occurs."

Hence the race. These days Craig's off on an ambitious series of deep-sea sampling trips using both surface vessels and submersibles. Rather than trying to identify everything, he and his crew focus on specific types of animals, such as marine worms, that are abundant enough to sample and study. "We're sampling over a broad region in the abyssal Pacific so we can get an idea of how much overlap there is as you move from one end of the area of mining interest to the other—a distance of about three thousand kilometers," he tells me. "The basic problem is that we don't know how broadly species are distributed in the deep sea. Whether you have the same fauna across this whole region or whether at one end you have one set of species and at the other you have another set of species. It looks like the latter is the case, in fact. If species are not completely distributed across the whole region, then you need to design MPA that recognize that at different places all along the areas of mining."

The line at the local recreation center finally ends at a counter, where I now stand face-to-face with a young woman sporting a ponytail and a nose stud. She peers intently at the computer screen that separates us as she tries valiantly to override a program that will not allow my six-year-old son to switch from the Tuesday/Thursday tadpole class he was in last term to the Wednesday/Friday pollywog class for this semester. Unsuccessful, she sighs, pulls out a paper form, and begins taking down my details, setting me on course to battle the bureaucracy that rules our city parks and recreation department. One child, one swim class, one pool in one city. I think of Craig and his ambition to map and protect the deep sea, the webs of bureaucracy—international, regional, federal, state—that govern our human interactions with the oceans. "A lot of the whale migration routes are within two hundred miles of the coast," he told me when we last spoke. "They are actually subject to the management of countries and the laws of countries. Whereas the management of the international seabed, and actually setting up MPAs in international waters, is quite a bit more complex. That's much more difficult to police, if you want to use that term; to manage, to enforce rules."

I am just trying to enroll my son in a swim class, and already the bureaucratic hoops feel daunting. Craig is trying to map a vast diversity of unknown species in the cold, dark, extraordinarily high-pressure aquatic conditions of the largest ecosystem on earth, an ecosystem that crosses a labyrinth of bureaucracies. I wish I could give him the powers of a superhero. A superhero for the deep.

18

Let's Talk about Sex, Baby

You can spot them a mile away—the mothers who don't get out to dinner very often. Faces flushed with excitement, bodies dressed in whatever was in style the year before they got pregnant, they teeter into a restaurant on unaccustomed high heels or scurry quickly to a seat, anxious to hide those "only-things-my-feet-will-still-fit-into" shoes beneath the table. I know these women. I am one of them.

But you wouldn't guess it tonight. Tonight even the stylish businessmen in the next booth are leaning toward our table. I am tempted to credit this unexpected attention to my new underwear—like many before me I have finally replaced my giant postmaternity underpants, or "Guppies," as my husband calls them, with sporty bikini briefs from Victoria's Secret—but that's clearly not it. The attraction is due purely to sex. Whale sex. "Right whales have one-ton testes," I proudly proclaim, my voice rising above the din of forks skipping across china. I am confident of my source. I have the goods straight from the mouth of Philip Clapham, self-acknowledged whale porn blogger to the world.

Philip didn't aspire to this highly coveted role. Indeed, he far prefers functioning in his professional capacity as a research scientist with the National Marine Mammal Laboratory at the Alaska Fisheries Science Center. It all started when he agreed to do the "ask the scientist" gig on the whale.net website. For a few weeks each year, he's the guy who will answer whatever marine mammal questions strike your fancy. A few years back, his stint happened to fall at the same time a juicy bit of whale sex trivia labeled "The real reason the sea is so salty" started making the e-mail rounds. According to the message, a blue whale produces over four hundred gallons of sperm when it ejaculates, but only 10 percent of that actually makes it to his mate. So 360 gallons are spilled into the ocean every time the whale unloads.

"I have no idea where it started," Phil says. "It's sometimes accompanied by a photo; I've had it sent to me dozens of times. It's clearly a Photoshopped image. It's supposed to be a blue whale hanging from a crane or something, and there's this enormous phallus below it. Actually it's not even a whale, it's a whale shark. It's not authentic."

And neither is the semen statistic. No one has ever seen blue whales mate, so we have no idea how efficient their sex act is. "Nature is usually very good at maximizing efficiency," says Phil, "so it's very unlikely that much of the semen enters the water rather than its intended target in the female anatomy. The thing that made me finally realize the gossip had gone global is that I got something from a so-called 'pavement engineer' whose name was Mohamed Ali in Pakistan, who said, "Excuse me, can you tell me if this is true?"

Whale penises, even fabricated ones, never fail to cause a stir. I still remember the screaming produced by a class full of fourth graders visiting the underwater viewing gallery at the Vancouver Aquarium. Hyak, a mature male killer whale, had just swum past the windows. He'd flashed them. Unlike the fellow lurking in the bushes at the local park, a whale doesn't have to work up to the event. "If you look at the structure of the penis in cetaceans, including large whales," says Phil "it is actually a fibroelastic penis, it's not a vascular penis, which means it doesn't fill with blood." Apparently antelope are built similarly. The penis is internal, but it comes out when it needs to. The very mobile tip of the penis reaches over and inserts itself into the female's vagina. "What you find in those is that copulation tends to be very brief," he explains. "It's sort of like, whip it in, whip it out."

These "quickies" may explain why no one has ever seen a humpback whale "do it." Humpback whales have been studied from one end of the earth to the other, but not one mating has ever been sighted. "We have logged millions of hours of observation of humpback whales in breeding grounds, and we certainly see them do everything else," Phil says. "We've seen males kick the crap out of each other in competitive groups. And of course there is the whole song issue, which is related one way or another to mating. But no one has ever seen it with humpbacks."

He groans when I mention a photo I'd seen in a book of two humpbacks mating during a spectacular aerial breach. "It's just ludicrous," he grumbles, referring to Lyall Watson's book on whales. "He has this picture in there of two humpback whales coming out of the water, and they're almost belly to belly and he has it labeled as humpback whales mating." Phil attributes some of the confusion to a myth, first proposed by old whalers, that humpbacks mated by jumping out of the water. "What they were almost certainly seeing, if they were seeing anything," he explains, "was males fighting and coming up and butting heads, which is what they do in the breeding grounds sometimes. It isn't exactly easy to copulate when you are jumping up in the air."

Hang on a second. If no one has ever seen a humpback mate, how do they know these are breeding grounds? Essentially, Phil tells me, they have seen everything but the act itself. Doc White and Ceci Devereaux have witnessed the mating ritual firsthand. "With the humpies, it's *really* violent," says Doc. "It was like a freight train. There were eight to twelve males chasing a male and a female, and I think, wow, that's sex." Phil concurs. "Even though we don't see copulation," he says, "we're 99.9 percent sure that's where it's going on." And then, of course, there are the calves. The gestation period of humpbacks is about a year. "Newborn calves are seen in profusion in the breeding grounds in the middle of winter," says Phil, "so obviously, the copulation is going on the year before."

You can't talk about whale sex with a group of mothers, I discover, and not end up talking about ovulation. There's always someone in the group trying to conceive, and any additional information, whatever the source, or the species relevance, seems welcomed. Some female humpbacks, it turns out, are polyestrous; they will ovulate more than once over the course of the winter. It's not clear what percentage of female humpbacks have more than one estrus in a year, or whether other whale species have the same capability. The general assumption is that they ovulate, and if they don't succeed in conceiving, they stick around and ovulate again.

"Do they have a clitoris?" my friend Sherry asks. "Phil says they do," I tell her. "And he says that none of the males can find it." With that, the fellows beside us straighten their backs and pretend to focus on their dinners. But I'm not fooled. "Phil's seen two male right whales copulating with the same female at the same time," I whisper. "One male will lie on each side and the tip of each penis enters the female at the same time." Voilà. The guys next door are back in the game. "The curious thing about right whales," Phil says, "is that you see them having sex year-round. Twelve months a year. But the calves are only born in midwinter following a gestation period of a year. Theoretically, mating is taking place the previous winter." In other words, they don't have calves year-round, just sex. "Why shouldn't they?" Phil laughs. "It may be that the females are kind of assessing males, you know, in a fun way, and then when it actually comes down to the real mating leading to conception, they've sorted out things. This is pure speculation. No one really knows why they do it, but they do it year-round."

Phil knows that he will never be free of questions about whale sex. They're just too darn fascinating. But in a perhaps misguided attempt to rid the world—and his in-box—of the dreaded blue whale cum message, he composed a brief article righting the misconceptions about whale sex. He begged his colleagues to circulate it widely, which they did, and of course it generated more questions. Right whales figure prominently in the piece, which is no surprise, given their rather "healthy" attributes. As the following excerpt reveals, right whales are the undisputed "he-men" of the whale world:

Right whales are large, rather rotund animals that live in both hemispheres. There are three species, two of which (the North Atlantic and North Pacific rights) are, by the way, critically endangered thanks largely to whaling.

A large male right whale would be around 17 meters (55 feet) and would weigh about 75 tons—much smaller than a big blue. So how much do a right whale's testes weigh? OK, get ready for this. The combined weight of a right whale's two testes is . . . one ton. Yep, ONE TON! As Dave Barry the humor columnist would say, we are not making this up.

OK, OK. I know what all you guys want to know next (and probably some of you women too), so to save myself from endless e-mail enquiries, here's the parallel statistic. Right whales also have a longer penis than any other whale, or indeed anything else on earth. The average length of a right whale's member is around 2.3 meters (7.5 feet), or about 14 percent of its body length. And in case you're wondering, both penis and testes are internal (guys, you wouldn't want all that additional drag if you spent your life moving through the water, now would you?) But the penis can be extruded outside the body for mating.

So there you go. Right whales have such huge testes because their mating system is based on what's called sperm competition. Female right whales will mate serially with multiple males. Consequently, each male tries to produce a huge volume of sperm to out-compete that of the male who was, er, there, before him.[1]

Right whales have one of the smallest brain-to-body-size ratios of any cetacean and the largest testes in the animal kingdom. As Jon Lien, marine mammal scientist at Memorial University in Newfoundland, famously phrased it, "If they think about anything, we know what they're thinking about."

You may be forgiven for not being familiar with the concept of sperm competition. Despite the fact that "billions of animals" do it, according to Phil, it draws a blank with my circle of friends. I guess the mating habits of dragonflies and ungulates just aren't that richly understood. There are even a few British scholars at the University of Manchester who wrote a series of articles claiming that we humans do it too.[2] "If you look at humpback whales by contrast," Phil says, "you'll find they have much smaller testes per body weight, and they also fight very aggressively. They are competing at the individual level, unlike right whales, which don't get into really aggressive contests with each other, but they compete at the level of sperm."

Female whales, in general, seem to have a lot of choice with respect to mating partners, even when it comes to the really big boys. Male sperm whales are enormous, one and a half times the length and three times the mass of a mature female. They are among the most extreme cases of sexual dimorphism, or differences in body size between the sexes, of any other mammal. People used to think the males were "harem masters," controlling a social group for long periods. In fact, their mating strategy is quite different. Male sperm whales travel between female groups searching for receptive females and staying with them for only a few hours at a time. But why are the males *so* huge? Hal Whitehead thinks it may be to impress the ladies.

"What seems to be happening is that it is not the great male who figures out who he wants to breed with, it's the females," explains Hal. "Sometimes the females think that the male who's arrived is great. You can see they get all excited, they make lots of codas as they come rushing over to him. Other times the male comes in making the big clicks, he struts in, and all the females say, forget it, they all dive and then he slinks off. It looks as though female choice has a fair amount to say about who breeds and about why they are so big. They are so big because perhaps the females like big males."[3]

Most surprising to Jonathan Gordon, a research fellow at the Gatty Marine Research Institute at the University of St. Andrews, was the shock and spectacle of seeing thirty "little" female whales, each about thirty tons, actively swarming an enormous male. "I had expected these huge males to be forcing their attentions on unwilling females," he writes. "What I observed underwater could not have been more different. The male was the focus of intense attention from all group members, who crowded in on him, rolling themselves along his huge body. They just seemed delighted that he was there. For his part the male was all calm serenity and gentleness. Even the calves were interested, and on one occasion we saw a male gently carrying a calf in its mouth."[4] Reflecting on the experience during a radio interview he adds, "It's the sort of thing you expect insects to do when [they go] sex-mad with pheromones. Not the thing you expect from complicated creatures like whales."[5]

The waiter appears with a list of dessert specials. He says something about a crème brûlée and a passion fruit bread pudding, but none of us pays attention to the details. We do, however, ask lots of questions. We want to hear him talk. He's got a honey voice that floats around the table and curls up on our laps. When he finally escapes under the pretext of giving us a moment to think, Meg lets out a gasp. "That voice!" she purrs. "I bet he's great in bed."

Female killer whales prefer males with distinctive voices too. Dr. Lance Barrett-Lennard at the Vancouver Aquarium Marine Science Centre has found a clear link between the calls that killer whales make and who they mate with.[6] A calf learns its dialect from its mother and other closely related adults, retains it for life, and passes it on to the next generation with few modifications. The more similar the dialects between two killer whale groups, the more closely related they are. Females apparently choose mates that don't sound like them and, by doing so, prevent inbreeding.

Somehow, inevitably, the conversation slides back to penises, or "pink floyds," as Ceci prefers to call them. The fact that gray whales mate in these very dense concentrations in the coastal lagoons off Baja makes it relatively easy for humans to observe. Sandy Lanham, the pilot who flies aerial surveys over the breeding lagoons, likens copulating gray whales to bunnies. "When you're flying over, it looks like the agitate cycle of your washing machine when you lift the lid," she says. Like Sandy, Ceci's seen loads of gray whale penises, and she has a Southern woman's gift for description. "I mean they are *large* bulbous things," she exclaims. "It almost reminded me of a tube worm . . . in larger scale. But yeah, nice and pink."

"I've actually touched a dolphin's penis," my buddy Georgia blurts out. "Ick. It actually rubbed against me. It was horrible." Apparently the beast shoved her hard against the side of a pool and wouldn't back off. It happened several years ago, when swim-with dolphin programs were just getting going. "Male bottlenose dolphins will molest anything with their penises, as many women who have been in swim-with dolphin programs will attest," Phil says. "They actually don't do this now, a lot of the time, with male dolphins, because they end up sexually harassing women. It's always women, not men, interestingly enough." He tells me about an e-mail he received from a woman in Russia. Apparently, you can arrange to go swimming with dolphins in the

Moscow Aquarium. "She wrote me this lovely e-mail in broken English explaining that she'd been swimming with dolphins and this dolphin had molested her. The thing that I loved about this e-mail was she said, 'If I do this again, should I be encouraging this behaviour?' I thought, I don't know, did you have a good time?"

Doc and Ceci have swum with wild dolphins and shared the experience with many others. They've noticed that dolphins not only seem able to differentiate between men and women, but they're also good at telling real women from those who, shall we say, have been enhanced. "We were in the water for twelve days with dolphins in the Bahama Banks." Ceci tells me, "Without exception they never went close to this one girl. She was really nice, very friendly. I think it was because she had a boob job. They couldn't understand it." Doc agrees. "With sonar that could easily be the case," he says. "With sonar, they would see that. They can even use it to tell if other dolphins are pregnant."

Can dolphins really "read" each other's reproductive status with their sonar, I wonder? It's such an intriguing idea, I run it by Sam Ridgway, a senior scientist at the U.S. Navy Marine Mammal Program and a professor at UC San Diego. "No evidence that I know of," he responds cheerfully.

When it comes to cetaceans and sex, however, evidence is clearly hard to come by and harder still to agree on. Recently, the question of dolphin homosexuality has been gaining attention in both the research literature and popular books.[7] "Are there gay dolphins?" I ask Phil. "It depends on how you define it," he responds. "Older males will do things sexually to younger males. It may just be a dominance thing rather than what we would think of as homosexuality in the sense of consenting partners both having a good time."

Bruce Bagemihl, author of *Biological Exuberance: Animal Homosexuality and Natural Diversity,* is quite familiar with the euphemisms researchers use to explain why animals might appear to engage in same-sex acts. It's dominance. It's a greeting. It's just playing. It's a contest of stamina. It's barter for food. It's aggression. Zoology, he argues, adheres to a folk model of homosexuality as perverse, unnatural, and bad and is far behind the humanities in recognizing it as a legitimate subject of inquiry. In fall 2006 the Oslo Mu-

seum of Natural History opened the first-ever museum exhibition dedicated
to gay animals. Geir Soeli, the project leader of the exhibition, stated, "Homosexuality has been observed for more than fifteen hundred animal species,
and is well documented for five hundred of them."[8] With excellent cover-
age from the international press and the museum experiencing robust
crowds, there's no question the debate will flourish.

Whether it's the wine or the fact that most of us are still getting up in
the night with our kids, the energy around the table is starting to wane. Unwilling to let go of an evening of freedom, I pull out a printout of a cus-
tomer review of Bagemihl's book I got off Amazon and begin to read it in
a clear, bright voice. "Dolphins have some tricks I'd never heard of. They
have 'nasal sex'—the insertion and stimulation of the penis in the blowhole;
and 'sonic sex'—the stimulation of the genitals using sonic pulses; as well
as 'beak-genital propulsion'—when the nose is inserted into the male or female genital slit, manually stimulating the genitals while propelling them
along." I pause to catch my breath and realize that no one at the table is
talking. "I've got a whole section on tool use and masturbation. Shall I go
on?" I ask. "Yes. Please," shouts a voice I do not recognize. I wait a moment
for him to scramble into the chair beside Meg. There. It's proven. Women
who don't get out to dinner very often are magnets for men. All it takes is
a little whale sex.

19

Missing Meat

It's a FedEx delivery nightmare. A piece of flesh from one of the world's rarest whales is lost in shipping. Luckily it's on ice. But its destination, Washington, DC, is in full summer swelter. If the flesh thaws, valuable genetic data will be lost. And whatever room the package lands in will stink to high heaven. As anyone who has ever tried to retrieve a lost package from FedEx knows, there is a vast thicket of bureaucracy to be navigated, and even then, the best you can hope for is the voice of a kindhearted but disempowered clerk speaking to you from some routing office in Kansas. The case is hopeless.

Enter an unlikely protagonist. Like a stealth bomber, he is quiet and inconspicuous, old jeans and tired T-shirt draped loosely across his thin frame. But Robert (Bob) Brownell, senior scientist for the Protected Resources Division at the Southwest Fisheries Science Center, the research arm of NOAA's National Marine Fisheries Service, knows the full value of that package. And bureaucracies? Hell, he's been battling that particular brand of evil his entire career. I find him in his office tucked between skyscrapers of research papers and faded conference proceedings, firing off e-mails and leaving voicemails for everyone who knows anyone who might be able to pass along a message. The message? "Please put the meat in the freezer."

What, exactly, is the value of a piece of meat? At the French Laundry in the Napa Valley—one of the most exclusive restaurants in the United States—a slice of Kobe beef demands an $80 upgrade above the $150 prix-fixe menu. A single serving of bluefin tuna sushi goes for $75 in finer Japanese restaurants. But for conservation biologists like Bob, a piece of western gray whale meat is priceless. In 2006 the total population of western gray whales, which were thought to be extinct as recently as 1972, was just 123 animals. This tiny population includes only twenty to twenty-five reproductively active females. A plug of western gray whale meat no bigger than a pencil eraser holds a multitude of secrets about the lives of these critically endangered creatures.

Western gray whales never recovered from the assault of twentieth-century whaling. The individuals who exist today continue to fall prey to ship collisions. Four females died in fishing gear entanglements off the Pacific Coast of Japan in the last two years.[1] Such waste makes Bob see red. "The fisheries agency of Japan is my biggest headache," he growls. "They have no conservation ethic. They support industry more than any other government that I've worked with, and they have no concerns whatsoever about conservation."

Yet the problems facing western gray whales don't end there. Their only known feeding ground lies along the northeastern coast of a remote Russian island, called Sakhalin, 6,200 miles east of Moscow. The trouble is, that faraway place is now smack in the epicenter of an international effort led by Sakhalin Energy to develop vast oil and gas reserves locked beneath the seafloor.[2]

None of the negotiators was thinking of gray whales when they hammered out the groundbreaking agreement in 1994 that called for multinational oil companies to provide the capital and technology to develop Sakhalin's energy fields, in return for allowing foreign companies access to the reserves and a split of the profits. When Bob blew the whistle about the critical location of these endangered whales' feeding grounds, there was a stunned reaction from Moscow, Washington, and oil headquarters throughout the world. "You have to be kidding" was the response Bob got.

"The biggest thing," says Bob a decade later, as he prepares for an International Whaling Commission (IWC) meeting, "is the impact of seismic surveys for oil and gas development on the whales." Bob leads the U.S side of the joint Russian-American research on the western gray whale and has been a member of the U.S. delegation to the IWC for thirty-plus years. He hands me a copy of a scientific paper, what has become the key case study to be used in the meeting.[3] "We show that in 2001 when they were doing some seismic surveys, the animals got displaced off the feeding grounds," he says. "The guys from Exxon said to me, 'Well, they were still feeding.' I said, 'Well, yes they were. But were they feeding as much as they were feeding before they were displaced? And why, two days after you stopped the survey, did they go back to where they were feeding before they were displaced? You did something, they moved. You stopped it. They came back.' I mean, it's the clearest case, almost like an experiment."

The amount of food a western gray whale eats during its summer feeding season literally spells the difference between its life and its death. Between 1999 and 2001, forty-eight skinny individuals were observed after Sakhalin-2's platform, Molikpak, began operating only twelve miles from the whale's primary feeding grounds. Mature females, including those known to be pregnant or lactating, were among them.[4] "Our concern is especially for mothers and calves," Bob tells me. "There is extreme site fidelity. All females in the population always go back to the same place with their calves." Mothers, he says, get pretty skinny when they're lactating. They're on a knife's-edge balance between being able to get the calf to survive and not being able to. "The calf gets weaned the first time it comes to the feeding ground when it is seven or eight months old," Bob explains. "Then it has to have enough reserves to migrate south and come back again. It can't feed again until the following June!"

Phase 1 of Sakhalin-2 has been in production since 1999, producing oil for six months each year during ice-free conditions. Phase 2 is intended to allow production of oil and gas year-round, with production beginning in 2008. It will mean construction of two new offshore platforms, offshore and onshore pipelines, and onshore processing and exporting facility. Under the auspices of IUCN, the World Conservation Union, an independent international scientific review panel was established to evaluate scientific aspects of western gray whale conservation in the context of phase 2.

If whale research was a branch of oncology, western gray whales would surely be considered a case with little hope for survival. The scientific review panel found Sakhalin's activities threaten the whales and their habitat in a number of horrifying ways, including noise caused by high-intensity geophysical seismic surveying and drilling operations, increased ship and air traffic, oil spills, chemical contamination of the animal communities the whales feed on, and accumulation of toxic substances in the whales' tissues and organs. "The most precautionary approach would be to suspend present operations and delay further development of the oil and gas reserves in the vicinity of the gray whale feeding grounds off Sakhalin," the panel concluded, "especially the critical nearshore feeding ground that is used preferentially by mothers and calves."[5] William Eichbaum, World Wildlife Fund's (WWF) vice president for endangered spaces, puts it more bluntly: "The recommendations of the world's most respected whale scientists are clear and unequivocal," he says. "If Shell routes this pipeline right through the heart of the whales' feeding grounds, it is potentially condemning them to extinction. We must not let an offshore oil platform become the tombstone of the western gray whale."[6]

Western gray whales once covered a huge range, marine ecologist Boris Worm confirms. Today barely one hundred individuals remain. "The whole phenomenon of the death of a species that we really care about makes my skin crawl," he confides. "We actually *know* that it is happening. There is no doubt. But there is a ton of denial."

The full tragedy of this sad tale, however, is that it's even worse than it appears. When it comes to conservation, what happens in one part of the world is often surprisingly connected to what happens in another. From Sakhalin Island, Russia plans to become a player in the market for liquefied natural gas (LNG) in the Pacific Ocean. This gas-field-to-carrier project will be the largest in the world. If it moves ahead as planned, Russia will be shipping liquefied natural gas to a regasification plant along the west coast of Baja California by the summer of 2008.[7] Another LNG terminal slated for Puerto Libertad, in the Gulf of California, means that the sea of milk, one of the richest places in the world to see the greatest diversity of whales, is on the verge of becoming a shipping lane for highly explosive fuel—fuel destined to be consumed by those of us who live in the United States.

"They want to bring these enormous gas boats into the northern gulf through the midriff," Exequiel Ezcurra explains when we meet in his office at the San Diego Natural History Museum. As both the museum's provost and director of its scientific research division, Exequiel is an erudite philosopher and a prolific advocate for the preservation of biodiversity and the monitoring of global change. He's built a prestigious academic career and was appointed by President Vicente Fox in 2001 to direct the National Institute of Ecology (INE) for Mexico. He pulls out a nautical chart and points to a cluster of tightly packed islands. He rattles off the English translations of channel names such as "little hell" and "get out if you can" to illustrate the danger of the tidal current in this narrow section of the gulf. "This is Rassa Island, where the seabirds nest, and this is where the sperm whales come to feed," he says, sweeping his finger across the same tangle of islands. "Now imagine this area with the traffic of immense boats. It is mind-boggling. It is terrifying, really. Just to decompress gas and pump it to Arizona." The boats Exequiel is talking about are thousand-foot-long cryogenic tankers. LNG is a natural gas that has been cooled and condensed into a liquid. Occupying six hundred times less space than it does in its gaseous state allows it to be shipped from remote locations to markets. At the receipt terminal, LNG is unloaded and stored until it can be vaporized back into natural gas and moved via pipeline to customers. The unintended consequences of an explosion are profound. "That happened in Texas, unfortunately, and virtually wiped out a town," says Doc White. "If it was nesting season, you could wipe out every bird within five miles, in an inferno."

The United States has four LNG terminals, all built in the 1970s on the East Coast and the Gulf of Mexico. Attempts in recent years to build facilities on the West Coast, in California, have foundered, owing to community concerns about health and safety threats. "With LNG you've got to come to a coastline, and in populous areas where the gas needs are, you don't have coastal areas available to be developed," Darcel L. Hulse, the president and chief executive officer of Sempra LNG has said.[8] "We'd love to be able to put one in San Diego Bay or Los Angeles Harbor, but those are much more difficult to permit. Thank goodness for Mexico allowing us to do it."

Such comments fill Exequiel with despair. Societies with the capacity to use the best scientific information available should also be able to practice sustainability. They should be able to see themselves in a broader context and to demand accountability in the long term from their decision makers and leaders. Finally, they should be transparent, open to participation and to information. Like Bob, Exequiel abhors the failure of governments to take the science of conservation seriously. "After being in the Mexican government for five years," he explains, "I can tell you with certainty that high-ranking governmental officers never think scientifically. They never think beyond the scope of four to five years, and they are terrified of having society knowing things."

What impresses me about Exequiel and Bob, and about so many of the scientists featured in this book, is that they persevere in spite of their understanding of the huge odds stacked against them. Bruce Mate, director of a marine mammal program, for instance, is carefully tagging whales in the healthy eastern Pacific gray whale population to test new technologies before applying them to the western gray whales. "Sometimes you do pilot studies, in less dire circumstances," he explains, "because you want to be able to earn the right, the respect, of the scientific community and of the NGO community to go do something that is even more meaningful." For Bruce procuring quality data from western gray whales is crucial to the whales' survival. Data cuts through various layers and levels of interests and bureaucracies. "Nobody is out to kill these animals off. There isn't a malicious effort to do that. They'd rather respond to scientific information than emotional hysteria."

What worries Bruce most is that we will lose endangered whales through benign neglect if we aren't more proactive in seeking information. "It's not like trying to put a man on the moon," he says. "You just need some time and space and funds to unravel the issue down to its basics and examine it. I think the answers become apparent, and the people who are in a position to change human behavior are willing to do that. Some of these things are pretty easy."

Easy, I have come to realize, is a relative term for a man who has dedicated his advocacy to the field of science and his life to the most endangered of large whales. Apparently tackling oil giants pales in comparison to trying to save beaked whales from being blasted by military sonar. In 2003 scientists discovered lesions caused by nitrogen bubbles in fourteen beaked whales stranded in the Canary Islands after sonars were used in a major international naval exercise on September 24, 2002. It was the first evidence that whales can suffer the bends. According to Antonio Fernández, a veterinary pathologist from the University of Las Palmas de Gran Canaria, who examined the whales, the bubbles could have formed because the deep-diving whales, startled by the sonar, surfaced too quickly or changed their diving patterns. This would have caused the nitrogen accumulated in their tissues to come out of solution and create bubbles large enough to block arteries.[9]

"They're getting their chimes rung," Bruce says. "That sounds a little less brutal than saying they're hemorrhaging through their eyes, ears, and blowholes. They're in harm's way in certain places because of the navy's need to be able to use some of those systems to protect us from diesel-powered subs with quiet props. Do we want them to catch a terrorist before they drive up onto a shelf off New York or Washington, DC? Of course we do. But how do we get around that dichotomy for an animal with a physiology and anatomy that is sensitive to those levels of sound?"

In the face of such gloomy predicaments I find myself drawing some comfort from Jay Barlow's take on the state of the world's whales. Jay is the head of the Coastal Marine Mammal Program at the Southwest Fisheries Science Center of NOAA Fisheries and the man behind some of the largest international whale studies in the world. The point he wants to convey is that most populations of large whales have been recovering since the cessation of whaling. "'Save the whales' is an outdated notion," he tells me. "Trouble is, that slogan has been so popular for so long, NGOs still like to use it to raise funds. They're not terribly comfortable with the idea that large whales might no longer be endangered." It's not that Jay isn't sensitive to the plight of western gray whales, beaked whales, and right whales. It's just that he wants to make it clear that they are tragic exceptions. "We've got to get used to the idea that most large whales are safe from extinction and start refocusing and rededicating ourselves to current conservation issues."

Indeed, the focus of conservation efforts, according to many leading ecologists, shouldn't be individual species but rather the state of the planet as a whole. No ecosystem is free from human influence. Jane Lubchenco, distinguished professor of zoology at Oregon State University, and her colleagues first made this profound assertion in a 1997 *Science* article.[10] It has been echoing down the corridors of our scientific institutions ever since. "To study ecology without humans in the equation is meaningless," Boris Worm tells me. "Yet at scientific meetings, I still have to get up and say, 'You are talking about putting all these instruments in the ocean to understand why animals are going where they do, yet you have forgotten to put us, the largest predator, into the picture.' You can't make sense of the oceanography without taking into account the global impact of humans on the planet as a whole," he explains. "Unless we grasp that, we will keep going with whatever approaches we already have, and there will be a demise."

Though deeply frustrated by his colleagues' failure to change their research paradigms, Boris is also sympathetic. His mother was a psychotherapist, and he sees these all-too-human reactions as a classic response to feeling overwhelmed. When faced with the enormity of global problems, we intuitively strive to sustain the unsustainable, argues the environmental political theorist Ingolfur Blühdorn.[11] "What is depressing is when we are stuck," Boris says, "when we feel we can't do anything. That is as true in our personal lives as it is collectively. Sometimes it is so painful you want to run out and do something. But you get grants and you publish your papers and you don't do anything about it. It is like being in a marriage and just chugging along and suddenly realizing that it is not fine. You either wait until it falls apart, or you do something about it. You become an activist in your own life."

In the fight for the world's endangered whales and the ecosystems they depend on, Boris, Bob, Exequiel, and Bruce have chosen to be activists in their own lives, quietly going about the business of changing the world.

20

Shifting Scale

Systems. Links. Connections. In a world where ecologists spin threads between human activities, earth, and ocean systems on a planetary scale, Exequiel Ezcurra must surely be the master weaver. His tapestry is the Gulf of
California. For thirty years he has studied the intricacies of its regions, the
individual threads of its biologies and its politics. Now he's tackling the grand
picture: understanding the Gulf of California as one big ecosystem. "We have
to seriously start looking at things at the scale of regions or even the whole
earth," he tells me as we make our way past a fossilized plesiosaur at the San
Diego Natural History Museum. He leads me through a maze of scientific
labs on the way to the staff room. We make friendly chitchat as he prepares
a cup of tea, each of us making connections between his large extended family and that of my friend, Manny Ezcurra, the Monterey Bay Aquarium's
shark specialist. But what really makes me smile is my drink. He's put it in
a "noodly master" cup.

Noodly masters, for those of you not in the fold, are followers of the Church
of the Flying Spaghetti Monster (otherwise known as Pastafarianism). This
elaborate spoof gained prominence during the May 2005 Kansas State Board
of Education hearings to decide whether intelligent design should be taught
as science along with the theory of evolution. Based on the theory that a Flying Spaghetti Monster created the universe, the church uses scientific graphs
to demonstrate how global warming, earthquakes, hurricanes, and other natural disasters are a direct effect of the shrinking numbers of pirates since the
1800s. "If the Intelligent Design theory is not based on faith, but instead an-

other scientific theory, as is claimed," the official website states in an open letter to the board, "then you must also allow our theory to be taught, as it is also based on science, not on faith."[1]

It doesn't seem fair, somehow, that I should be here. There are prepubescent boys who would *kill* to be in spitting distance of seventy-million-year-old marine reptiles while chatting about an omnipotent being that shapes the world with its long pasta appendages—particularly one featured in the January 2006 issue of *Playboy*. But I'm glad it's me. Exequiel has had more than his fair share of Gulf of California adventures, and he's no stranger to pirates or high-action tales. "I was at the wheel, and everyone was asleep," he begins a story about a recent research expedition in the gulf. "The captain came down to check that everything was all right. It wasn't." A blip on the radar screen revealed that someone was following their boat. Exequiel turned off every light and peered into the darkness. He couldn't see a thing. Whoever was chasing them was coming with lights off, and they were catching up fast. "You know, you read so many stories about pirates in the seas," he says. "I wasn't happy." The real heart-in-the-throat moment came when the blip disappeared from the screen. Ten miles out to sea. Rough swells. Two o'clock in the morning. And a mystery boat so close, it could no longer be detected by radar.

"A huge beam of light suddenly floods the boat, and a voice screams: 'Identify yourself,'" Exequiel tells me. For a moment, he thought it was the navy patrol, trolling the water for drug-runners. But it wasn't. It was fishermen. "They patrol the coast themselves," he explains. "It's an amazing set of anarchist communities—ten or eleven fishing cooperates that run between Cedros Island and San Ignacio Lagoon. They want nothing to do with government. They have their own law enforcement. Many of them generate their own electricity. They have their own schools. They have their own airplanes to take the catch out to Ensenada."

They're also an impressive example of community management of a resource. Wild abalone used to be abundant all the way from Vancouver, Canada, to Los Cabos, Mexico, but not anymore. Abalone has been overfished along the entire coastline. Today a single pink abalone *(Haliotis corrugata)* sells for $32. And that's cheap, compared to the $100 an individual white abalone *(Haliotis sorenseni)* commands. The co-ops are the only place left where wild abalone is doing really well. "These guys got their authorization way back in the fifties or sixties, so the community owns one part of the coast or at least has a federal concession to that part of the coast," Exequiel says. There's no question that they use their minimal resources to maximum effect. "When they finally turned off the light and took off," says Exequiel, "I could see that these guys were riding a panga with an outboard engine. It was just a damn panga!"

Exequiel clearly relishes stories in which the "little guy"—or girl, as the case may be—rules victorious. "I have big admiration for 50 percent of humanity," he says as he warms his way into another story about a "wonderful, wonderful woman, a dear, dear friend of mine," named Enriqueta Velarde. "I always present her work as an example of how a single woman with no guns, no law enforcement, no weapons, can bring back two species from the brink of extinction, just by going to a single island every spring and interacting with the local fishermen, with the local people there."

For the past thirty-five years Enriqueta has been saving seabirds on Rasa Island, the tiniest island in the Gulf of California. When she first arrived, fishermen routinely went there to ransack the nests. They'd break every egg on the island to force the females to lay again. That way, when they returned to collect the eggs in three days' time, they could be sure they were all fresh. "It was illegal, a terrible waste, but it still happened, Exequiel says. "There were only five thousand birds when Enriqueta started going there, and now there are a million. This is almost 100 percent of all the elegant terns *(Sterna elegans)* and Heermann's gulls *(Larus heermanni)* in the world that nest there."

Exequiel analyzed twenty-five years of Enriqueta's data to better understand how this remarkable recovery occurred. Even after she was able to convince the fishers to stop destroying the eggs, there were still years in which the birds would come in and lay eggs, and the chicks would hatch and then die. The villain responsible for these deaths was not human, and it wasn't local. It originated thousands and thousands of miles away. The chicks perished during the El Niño years when the water in the gulf was really warm. "What makes this story intriguing," he says, "is how a change in atmospheric pressure between northern Australia and Tahiti, some fifteen to twenty thousand kilometers away can produce a chain of changes in the currents that will eventually accumulate hot water in the Gulf of California, which will bring down the population of pelagic fish. The female birds could lay because they came in from somewhere else with their bodies full of fat, but they couldn't feed their chicks because there wasn't enough ocean productivity. There weren't enough small fish."

Teleconnections—links between events that span distance and time—sounds like some new age concept, but Exequiel assures me that it's serious science. "A lot of phenomena that we think of as 'natural catastrophes' are actually predictable," he explains. The trick is finding a scale big enough in which to see them. The story of El Niños and the chicks is a perfect example. By using mathematical models based on the links between El Niños and seabird chicks, fishery managers can make advanced predictions about the sardine catch and adjust the effort of the fishing fleet accordingly.

The dividing line between the Gulf of California and the Pacific Ocean is at the southern tip of the Baja Peninsula near the lovely colonial town of San Jose del Cabo. A mosaic mural graces the entrance to the cathedral in the town's central plaza. Stand too close, and you can't see the picture it depicts. You need to take ten giant steps backward and then look at the tiles en masse to see the centuries-old tales of conflicts between the Jesuits and Indians. In art, as in ecosystems, shifting the scale is one of the best ways to see hidden connections. Charles Darwin knew this well. His diaries from his 1831 voyage on the *Beagle* are filled with examples of his looking at the beaks of finches and then making an inference that finches have different beaks because natural selection drives them to specialize to eat a certain type of fruit. Then he would generalize this inference and say, if this is true, then it must also hold true for other islands with other species. Darwin would develop a theory at a microscale and then generalize it by looking at a macroscale, and vice versa. He used change in scale as a way to test the robustness of his ideas.

A generation earlier, in the 1790s, Alexander von Humboldt developed the first ideas about continental drift using the same technique. Steinbeck and Ricketts paid homage to this approach on their own voyage a hundred and fifty years later. "Collecting large numbers of animals presents an entirely different aspect and makes one see an entirely different picture," Steinbeck wrote in *The Log from the Sea of Cortez*. "In a way, ours is the older method, somewhat like that of Darwin on the *Beagle*. He was called a 'naturalist.' He wanted to see everything, rocks and flora and fauna, marine and terrestrial. We came to envy this Darwin on his sailing ship. He had so much room and so much time. . . . We must have time to think and to look and to consider."[2]

These days, when scientists make so many breakthroughs by examining life at the molecular level in the controlled environment of a lab, Exequiel clearly sides with Darwin and Steinbeck. "These guys were traveling, seeing many places and developing really global hypotheses," he says. "They had the capacity to see the planet as a laboratory in itself." Exequiel is convinced that big problems, such as the extinction crisis or the degradation of earth's atmosphere, can't be solved solely on the basis of experimental design. You can't follow the typical experimental protocol and have one earth with greenhouse gases and another without in order to prove a point. "A lot of the discussion we have around the United States not signing the Kyoto Protocol has to do with that," he explains.

> They say, OK, you have to demonstrate to me experimentally beyond any reasonable doubt that greenhouse gases are indeed causing an increase in global temperature, otherwise the U.S. will do nothing about it. By definition it is impossible to demonstrate this, because you can't replicate the phenomenon you are studying, and you cannot have a controlled treatment, which would be of course an atmosphere without greenhouse gases. You have to resort to all these things that a scientist of the nineteenth century or the eighteenth cen-

tury resorted to, which is making inferences, using scale, accepting rationalism as your main method for explaining reality, and using the comparative method.

The trouble with a world that conceives of science primarily in terms of experimental design is that it supports a kind of "if it can't be measured, it can't be real" perspective that ignores the scale and scope of complex environmental threats. As Ulrich Beck, the German sociologist, puts it, "with the past decisions on nuclear energy and our contemporary decisions on the use of genetic technology, human genetics, nanotechnology, computer sciences and so forth, we set off unpredictable, uncontrollable and incommunicable consequences that endanger life on earth."[3] Yet at the same time we perpetuate a belief that science and technology can safely calculate and nullify these risks. Ironically, the same science that helps us to understand connections between seabird chicks and El Niños is incapable of comprehending the extraordinary complexity of the environmental threats we impose on those systems.

When Exequiel contemplates the threat of LNG tankers navigating past Rasa Island, he sometimes feels hopeless. "As an explorer, diver, naturalist, I have seen so many parts of the Baja Peninsula destroyed, for the ephemeral benefit of spring breakers and big hotel chains, that sometimes I get pessimistic about it," he says. "On the other hand, there are still a lot of things surviving that are wonderful, and we have to nurture some sort of optimism working in the environment, and some long-term vision."

Exequiel draws strength from fellow naturalists who, like him, have championed big picture ecosystem perspectives at different points in history. Darwin had no taste for public controversy, yet he published his theory of evolution by natural selection knowing full well that it was bound to offend the early Victorian public in England; after all, his theory was even unpopular with his wife. Steinbeck attempted to present, with the aid of science, an ecosystems view of our place in the world—no matter whose prejudices might be destroyed as a consequence. As early as the 1940s, his subject matter was drawing attention to the degradation of the environment. His themes of holistic ecology and conservation, ideas that drive ecologists and environmental groups today, were way before his time. Many critics slammed Steinbeck for equating people with animals, for creating a holistic hierarchy of the natural world. "Steinbeck's heroes are always the ones connected to the natural environment; his troubled characters are frequently disassociated or ignorant of being part of a larger whole," explains Brian Railsback, a Steinbeck scholar and professor at Western Carolina University. "Doc from *Cannery Row* and Tom Joad from *The Grapes of Wrath,* for instance, have this understanding of their roles in the big picture. It's what defines them as characters and makes them wise."[4]

Big-picture thinking is the really big news in biology today. At the 2007 TED (Technology, Entertainment, Design) Conference, an invitation-only event where the world's leading thinkers and doers gather to find inspiration, world-renowned biologist E. O. Wilson was awarded $100,000 and the support of the attendees to grant a wish that would change the world.[5] He chose to create a virtual "Encyclopedia of Life," a collaborative global Web-based project designed to catalog every living species in a flexible reference system.[6] The man dubbed "Darwin's natural heir" by the *Guardian* newspaper in recognition of his extraordinary contributions to modern scientific thought, Wilson is not alone in viewing the virtual world as the source of salvation for earth's increasingly threatened living ecosystems.[7] In the not-so-distant future, platforms already underway, such as "Google Ocean" and "Digital Earth," will enable us to observe the interconnected nature of planetary forces and global currents; to sail through a virtual ocean of scenarios too vast for us to currently comprehend; and thus, it is hoped, to make wiser, better-informed decisions. "Our next challenge as a species, for us to survive on this planet," says Tierney Thys, biologist and documentary filmmaker, "is to find our place in the matrix of life, as we create the matrix itself."

Steinbeck understood the power of such a vision. We "are related to the whole thing, related inextricably to all reality, known and unknowable," he wrote in the *Log from the Sea of Cortez.* "The profound feeling of it made a

Jesus, a St. Augustine, a St. Francis, a Roger Bacon, a Charles Darwin, and an Einstein. Each of them in his own tempo and with his own voice discovered and reaffirmed with astonishment the knowledge that all things are one thing and that one thing is all things—plankton, a shimmering phosphorescence on the sea and the spinning planets and an expanding universe, all bound together by the elastic string of time. It is advisable to look from the tide pool to the stars and then back to the tide pool again."[8]

When Lori Marino plunged a mirror into a dolphin exhibit, she discovered an animal that was self-aware. When I look in a mirror, I too am self-aware. Perhaps too self-aware. I see only myself. Yet lately I have been training myself to look at the other objects reflected in the mirror: the chair I am sitting in, the paint color on the wall, a stream of sunlight through the window. That is the challenge for me. I need to push my mind to put the diversity of life back into my field of vision. Whales help me to do that. They tantalize me with the knowledge that there are other fascinating, highly developed cultures inhabiting the earth and its oceans. The enormity of their presence cannot be overlooked. Whales inspire me to contemplate connections. They inspire me to act more generously. They inspire me to experience life in whale scale.

Introduction

1. E. Hoyt, *Whale Watching 2001: Worldwide Tourism Numbers, Expenditures, and Expanding Socioeconomic Benefits* (Yarmouth Port, MA: International Fund for Animal Welfare, 2001), 158; M. Honey, "The Business of EcoTourism," *GreenMoneyJournal.com.* Available at www.greenmoneyjournal.com/article .mpl?articleid=467&newsletterid=4.

1. Extreme Motherhood

1. O. T. Oftedal, "Lactation in Whales and Dolphins: Evidence of Divergence between Baleen- and Toothed-Species," *Journal of Mammary Gland Biology and Neoplasia* 2 (1997): 205–30.

2. R. Ternullo and N. Black, "Predation Behavior of Transient Killer Whales in Monterey Bay, California" (proceedings of the Fourth International Orca Symposium and Workshop, Chize, France, September 23–28, 2002). Available at www.montereybaywhalewatch.com/Features/KillerWhalePredation0210 .htm (accessed September 22, 2007).

2. A Sea of Milk

1. W. Sears and M. Sears, *The Baby Book* (New York: Little Brown, 2003).

2. O. T. Oftedal, "Animal Nutrition and Metabolism Group Symposium on 'Regulation of Maternal Reserves and Effects on Lactation and the Nutrition of Young Animals': Use of Maternal Reserves as a Lactation Strategy in Large Mammals," *Proceedings of the Nutrition Society* 59 (2000): 99–106.

3. Blue whales were greatly overexploited by commercial whalers in the nine-teenth and twentieth centuries. In 1960 the International Whaling Commission (IWC) banned the taking of blue whales in the North Atlantic, banned it entirely in the North Pacific in 1966, and banned it in the southern hemisphere in 1967. However, illegal unreported and unregulated hunting may have continued until the early 1970s. In 1985–86 the IWC declared a global moratorium on commercial whaling. For more on this issue, see Australian Department of the Environment and Heritage, "Blue, Fin and Sei Whale Recovery Plan 2005–2010," available at www.environment.gov.au/biodiversity/threatened/publications/recovery/balaenoptera-sp/index.html.

4. The Oregon State Marine Mammal Program hosts an excellent website describing current blue whale tagging and census research: http://oregonstate.edu/groups/marinemammal/NewBlue.htm.

5. The TOPP research project explores the Pacific Ocean, using a carefully selected group of animals from its ecosystems to gather data about their world. As members of this pilot program of the Census of Marine Life (COML), an international endeavor to determine what lives, has lived, and will live in the world's oceans, TOPP scientists will tag individuals from twenty-one species of marine predators in the Eastern Pacific to obtain an "organism's-eye" view of their world. Jointly run by Stanford's Hopkins Marine Lab, the University of California, Santa Cruz's Long Marine Laboratory, NOAA's Pacific Fisheries Ecosystems Lab, and the Monterey Bay Aquarium, TOPP also includes team members from several countries. Their website is www.toppcensus.org.

4. Resident Aliens?

1. For trends in American purchases of Mexican property, please see www.realtor.org.

2. M. Bérubé, J. Urbán, A. E. Dizon, R. L. Brownell, and P. J. Palsbøll, "Genetic Identification of a Small and Highly Isolated Population of Fin Whales (*Balaenoptera physalus*) in the Sea of Cortez, México," *Conservation Genetics* 3, no. 2 (2002): 183–90.

3. D. Harman, "More Americans Boost Baja Real Estate Boom," *USA Today*, November 30, 2005.

4. H. Ajami et al., *Alternative Futures for the Region of Loreto, Baja California Sur,*

Mexico (Cambridge, MA: Harvard University, 2005). Available at www.futur
osalternativosloreto.org/report/LoretoReport.pdf (accessed September 22,
2007).

5. How to Make a Really Rich Sea

1. John Steinbeck, *Working Days: The Journals of the Grapes of Wrath, 1938–1941,*
   ed. Robert J. Demott (New York: Viking, 1989).

2. B. E. Railsback, *Parallel Expeditions: Charles Darwin and the Art of John Steinbeck* (Moscow: University of Idaho Press, 1995), 13.

3. Steinbeck, *The Log from the Sea of Cortez* (1951; repr., New York: Penguin,
   1995).

4. Steinbeck, *Log from the Sea of Cortez,* 71–72.

6. Popular Mechanics

1. For an overview of human evolution, see the Smithsonian Institution's Human
   Origins Program website: "The Origin of the Genus *Homo,*" www.mnh.si
   .edu/anthro/humanorigins/faq/Encarta/genushomo.htm.

2. J. Goodall, "Tool-Using and Aimed Throwing in a Community of Free-Living
   Chimpanzees," *Nature* 201 (1964): 1264–66; the Jane Goodall Institute provides an overview of chimpanzee tool use on its website: "Jane Goodall: Study
   Corner: Tool Use," www.janegoodall.org/jane/study-corner/chimpanzees/
   tool-use.asp.

3. A. Whiten et al., "Cultures in Chimpanzees," *Nature* 399 (1999): 682–85.

4. To view a video of a chimpanzee making and using a spear, see National
   Geographic, "Video: Chimps Make and Use 'Spears' to Hunt," *National
   Geographic News,* February 22, 2007, http://news.nationalgeographic.com/
   news/2007/02/070222-chimp-video.html.

5. J. Mercader, M. Panger, and C. Boesch, "Excavation of a Chimpanzee Stone
   Tool Site in the African Rainforest," *Science* 296 (2002): 1452–55.

6. For more on chimp-made hammers, see University of Calgary, "Ancient
   Chimp-Made 'Hammers' Fuel Evolutionary Debate," *University of Calgary News and Events,* February 12, 2007, www.ucalgary.ca/news/feb2007/
   chimp-hammers.

7. F. B. M. de Waal, *The Ape and the Sushi Master: Cultural Reflections by a Primatologist* (New York: Basic Books, 2001); Frans de Waal, "Cultural Primatology Comes of Age," *Nature* 399 (1999): 635.

8. J. Marks, *What It Means to Be 98% Chimpanzee: Apes, People, and Their Genes* (Berkeley: University of California Press, 2002).

9. To view a photo of a dolphin using a sponge, see University of New South Wales Faculty of Science, "Dolphin Mothers Teach Their Kids to Sponge," *News Archive,* June 7, 2005, www2.science.unsw.edu.au/news/2005/dolphinspongers.html.

10. M. Kruetzen, J. Mann, M. Heithaus, R. Connor, L. Bejder, and B. Sherwin, "Cultural Transmission of Tool Use in Bottlenose Dolphins," *Proceedings of the National Academy of Sciences* 105, no. 25 (2005): 8939–43.

11. R. Stein, "'Sponging' Dolphins May Be Sharing Culture," *Washington Post,* June 27, 2005.

12. Adrienne Zihlman, quoted in R. Weiss, "For First Time, Chimps Seen Making Weapons for Hunting," *Washington Post,* February 23, 2007. Zihlman was awarded the 2004 Squeaky Wheel Award by the Committee on the Status of Women in Anthropology (COSWA). She has written extensively on gender differences in many aspects of physical anthropology, including morphology and locomotion, women in human origins, woman as gatherer; the life history of great apes, and the roots of sociality. Her continued efforts have been responsible for ensuring that an evolutionary perspective of humans includes women's contributions.

7. Mirror, Mirror on the Wall, Who's the Smartest of Them All?

1. M. Pendergrast, *Mirror Mirror: A History of the Human Love Affair with Reflection* (New York: Basic Books, 2003).

2. D. Reiss and L. Marino, "Self-Recognition in the Bottlenose Dolphin: A Case of Cognitive Convergence," *Proceedings of the National Academy of Sciences* 98, no. 10 (2001): 5937–42; D. Sarko, L. Marino, and D. Reiss, "A Bottlenose Dolphin's (*Tursiops truncatus*) Responses to Its Mirror Image: Further Analysis," *International Journal of Comparative Psychology* 15 (2003): 69–76.

3. V. M. Janik, "Whistle Matching in Wild Bottlenose Dolphins *(Tursiops truncatus),*" *Science* 289 (2000): 1355–57.

4. "Dolphins 'Have Their Own Names,'" *BBC News,* May 8, 2006, http://news.bbc.co.uk/2/hi/uk_news/scotland/edinburgh_and_east/4750471.stm (accessed September 24, 2007).

5. J. D. Smith, W. E. Shields, and D. A. Washburn, "A Comparative Approach to Metacognition and Uncertainty Monitoring," *Behavioral and Brain Sciences* 26 (2003): 317–39.

6. L. Marino, "Convergence of Complex Cognitive Abilities in Cetaceans and Primates," *Brain, Behavior and Evolution* 59 (2002): 21–32.

7. For a Renaissance illustration of the Great Chain of Being, see www.stanford.edu/class/engl174b/chain.html.

8. P. R. Hof and E. van der Gucht, "Structure of the Cerebral Cortex of the Humpback Whale, *Megaptera novaeangliae* (Cetacea, Mysticeti, Balaenopteridae)," *Anatomical Record: Advances in Integrative Anatomy and Evolutionary Biology* 290, no. 1 (2007): 1–31.

9. P. R. Hof, quoted in A. Coghlan, "Whales Boast the Brain Cells That 'Make Us Human,'" *New Scientist* (Nov. 2006), www.newscientist.com/article/dn10661-whales-boast-the-brain-cells-that-make-us-human.html (accessed September 24, 2007).

8. Building Nets from Bubbles and Other Mysterious Humpback Whale Talents

1. A. Darby, "DNA of Protected Whales Found at Japanese Market," *Sydney Morning Herald,* July 20, 2004, www.smh.com.au/articles/2004/07/19/1090089101723.html (accessed September 24, 2007).

2. C. S. Baker, G. M. Lento, F. Cipriano, M. L. Dalebout, and S. R. Palumbi, "Scientific Whaling: Source of Illegal Products for Market?" *Science* 290 (2000): 1695–96; C. S. Baker, G. M. Lento, F. Cipriano, and S. R. Palumbi, "Predicted Decline of Protected Whales Based on Molecular Genetic Monitoring of Japanese and Korean Markets," *Proceedings: Biological Sciences* 267, no. 1449 (2000): 1191–99; C. S. Baker, M. L. Dalebout, G. M. Lento, and N. Funahashi, "Gray Whale Products Sold in Commercial Markets along the Pacific Coast of Japan," *Marine Mammal Science* 18 (2002): 295–300.

9. Do Baby Sperm Whales Suck Milk through Their Noses?

1. S. Gero and H. Whitehead, "Suckling Behavior in Sperm Whale Calves: Observations and Hypotheses," *Marine Mammal Science* 23 (2007): 398–413.

10. Deep Culture

1. For a comparison of brain weights, see E. H. Chudler, "Brain Facts and Figures," http://faculty.washington.edu/chudler/facts.html.

2. For more on the evolution of killer whale species, see Government of Canada, "Whale Fin ID," *Science and Technology for Canadians,* www.science.gc.ca/default.asp?Lang=En&n=AB95B43D-1&edit=off.

3. For more on the evolution of killer whale species in British Columbia, see http://www.cascadiaresearch.org/robin/Oecologia1992.pdf

4. H. Whitehead, *Sperm Whales: Social Evolution in the Ocean* (Chicago: University of Chicago Press, 2003), 318.

5. W. Gilly et al., "Vertical and Horizontal Migrations by the Jumbo Squid *Dosidicus gigas* Revealed by Electronic Tagging," *Marine Ecology Progress Series* 324 (2006): 1–17.

6. For more on the satellite tagging of squid, please go to www.topp.org.

7. You can follow satellite tagged squid in real time at www.topp.org.

8. "Real Sea Monsters," *Monterey County Weekly,* March 10, 2005. Available at www.montereycountyweekly.com/issues/Issue.03-10-2005/cover/Article.cover_story.

9. B. Worm et al., "Impacts of Biodiversity Loss on Ocean Ecosystem Services," *Science* 314 (2006): 787–90.

10. L. D. Zeidberg and B. H. Robison, "Invasive Range Expansion by the Humboldt Squid, *Dosidicus gigas,* in the Eastern North Pacific," *Proceedings of the National Academy of Sciences* 104, no. 31 (2007): 12,948–50.

11. Details on the 2006 Symposium on Fisheries Depredation by Killer and Sperm Whales may be found at www.killerwhale.org/depredation.

12. R. L. Parry and D. Bhat, "Dying Tribe Takes On Timber Giants over Lost Habitat," *Times,* May 5, 2006, www.timesonline.co.uk/tol/news/world/asia/article713387.ece (accessed April 23, 2008).

11. What's the Use of Granny?

1. J. Diamond, "Why Women Change," *Discover* 17, no. 7 (1996): 130–37.

2. J. Mann, R. Connor, P. Tyack, and H. Whitehead, eds., *Cetacean Societies: Field Studies of Dolphins and Whales* (Chicago: University of Chicago Press, 2000).

3. Diamond, "Why Women Change," 130–37.

4. K. McComb, C. Moss, S. M. Durant, L. Baker, and S. Sayialel, "Matriarchs as Repositories of Social Knowledge in African Elephants," *Science* 292 (2001): 491–94.

5. K. McComb, D. Reby, L. Baker, C. Moss, and S. Sayialel, "Long-Distance Communication of Cues to Social Identity in African Elephants," *Animal Behaviour* 65 (2003): 317–29.

12. Dolphin Snatchers

1. D. Weihs, "The Hydrodynamics of Dolphin Drafting," *Journal of Biology* 3, no. 8 (2004).

2. E. F. Edwards, "Behavioral Contributions to Separation and Subsequent Mortality of Dolphin Calves Chased by Tuna Purse-Seiners in the Eastern Tropical Pacific Ocean," *Southwest Fisheries Center Administrative Report LJ-02-28* (2002).

3. S. R. Noren and E. F. Edwards, "Physiological and Behavioral Development in Delphinid Calves: Implications for Calf Separation and Mortality Due to Tuna Purse-seine Sets," *Marine Mammal Science* 23, no. 1 (2007): 15–29.

4. D. Weihs, quoted in D. Adam, "Friendly Fishing Still Kills Dolphins," *Guardian,* May 4, 2004, www.guardian.co.uk/environment/2004/may/04/sciencenews.fish (accessed September 24, 2007).

5. Noren and Edwards, "Physiological and Behavioral Development," 15–29.

6. The International Dolphin Conservation Program Act Reports may be viewed at http://swfsc.noaa.gov/textblock.aspx?Division=PRD&ParentMenuId=228&id=5298.

7. An overview of the tuna-dolphin issue is available at NOAA Fisheries Service, "The Tuna-Dolphin Issue," *Southwest Fisheries Science Center,*

http://swfsc.noaa.gov/textblock.aspx?Division=PRD&ParentMenuId=
228&id=1408.

8. See NOAA Fisheries Service, "The Tuna-Dolphin Issue," http://swfsc
.noaa.gov/textblock.aspx?Division=PRD&ParentMenuId=228&id=1408.

9. T. Gerrodette and J. Forcada, "Non-recovery of Two Spotted and Spinner
Dolphin Populations in the Eastern Tropical Pacific Ocean," *Marine Ecology
Progress Series* 291 (2005): 1–21.

10. National Oceanic and Atmospheric Association, "Commerce Department
Issues Initial Finding on Tuna/Dolphin Interactions; Will Adopt New Dolphin-
Safe Label Standard" (press release, April 29, 1999), www.publicaffairs.noaa
.gov/releases99/apr99/noaa99–35.html.

11. Maureen McGowan, "'Dolphin-Safe' Tuna Label Safe for Now," *The National
Sea Grant Law Center,* www.olemiss.edu/orgs/SGLC/National/SandBar/3
.3tuna.htm#14 (accessed September 24, 2007).

12. K. A. Forney, D. J. St. Aubin, and S. J. Chivers, "Chase Encirclement Stress
Studies on Dolphins Involved in Eastern Tropical Pacific Ocean Purse-seine
Operations during 2001," *Southwest Fisheries Center Administrative Report*
LJ-02–32.27p, 2002; E. Santurtun-Oliveros and F. Galindo-Maldonado,
"Coping Behaviors of Spotted Dolphins during Fishing Sets," *Southwest Fish-
eries Center Administrative Report* LJ-02–36C (2002).

13. P. C. Perkins and E. F. Edwards, "Capture Rate as a Function of School Size
in Pantropical Spotted Dolphins, *Stenella attenuata,* in the Eastern Tropical
Pacific Ocean," *Fisheries Bulletin* 97 (1999): 542–54.

14. Santurtun-Oliveros and Galindo-Maldonado, "Coping Behaviors."

15. D. Lusseau and M. E. J. Newman, "Identifying the Role That Animals Play
in Their Social Networks," *Proceedings of the Royal Society of London* B 271:
S477–S481.

16. S. Vedantam, "A Political Debate on Stress Disorder: As Claims Rise, VA
Takes Stock," *Washington Post,* December 27, 2005.

17. G. A. Bradshaw, A. N. Schore, J. L. Brown, J. H. Poole, and C. J. Moss, "Elephant Breakdown," *Nature* 433 (2005): 807.

18. The Earth Island Institute operates an international monitoring program that
lists dolphin-safe fishing and processing companies worldwide, available at
www.earthisland.org/dolphinSafeTuna.

13. Friendly Mothers, Friendly Calves?

1. A. M. Kuris, A. R. Blaustein, and J. J. Alio, "Hosts as Islands," *American Naturalist* 116, no. 4 (1980): 570–86.

2. C. M. Callahan and B. S. Arbogast, "Biogeography and Molecular Systematics of Whale Lice Living on Gray Whale Islands" (symposium conducted at the Nineteenth Annual California State University Biotechnology Symposium, Los Angeles, 2007); Callahan and Arbogast, "Phylogeography of Whale Lice (Cyamidae) Living on Gray Whale Islands" (symposium conducted at the Eighty-sixth Annual Meeting of the American Society of Mammalogists, Amherst, University of Massachusetts, 2006).

3. J. Roman and S. R. Palumbi, "Whales before Whaling in the North Atlantic," *Science* 301 (2003), 508–10.

4. M. Shwartz, "Whale Populations Are Too Low to Resume Commercial Hunting, Geneticists Find," *Stanford Report,* July 24, 2003, http://news-service.stanford.edu/news/2003/august6/whales-86.html (accessed September 24, 2007).

5. J. M. Grebmeier et al., "A Major Ecosystem Shift in the Northern Bering Sea," *Science* 311 (2006): 1461–64.

14. The War on Fish

1. R. A. Myers and B. Worm, "Rapid Worldwide Depletion of Predatory Fish Communities," *Nature* 423 (2003): 280–83.

2. B. Worm et al., "Impacts of Biodiversity Loss on Ocean Ecosystem Services," *Science* 314 (2006): 787–90.

3. D. Pauly, "Depletion of World Fish Stocks," *The Science Show* (ABC Radio National, October 21, 2006). Available at www.abc.net.au/rn/scienceshow/stories/2006/1766715.htm (accessed September 24, 2007).

4. P. K. Dayton, "Reversal of the Burden of Proof in Fisheries Management," *Science* 279 (1998): 821–22.

5. W. K. Stevens, "Man Moves Down the Marine Food Chain, Creating Havoc," *New York Times,* February 10, 1998.

6. M. L. Cruz-Torres, "'Pink Gold Rush': Shrimp Aquaculture, Sustainable Development and the Environment in Northwestern Mexico," *Journal of Political Ecology* 7 (2001): 63–90.

15. Why Blue Whales Gotta Be Big

1. P. J. Jarman, "The Social Organisation of Antelope in Relation to Their Ecology," *Behaviour* 48 (1974): 215–67.

16. What You Can See by Listening

1. D. Brand, "Christopher Clark: Whales off Newfoundland Can Hear Whales near Bermuda," *Cornell Chronicle,* February 24, 2005. Available at www.news .cornell.edu/Chronicle/05/2.24.05/AAAS.Clark.whales.html (accessed September 24, 2007).

17. What You Can Learn from the Dead

1. G. W. Rouse, S. K. Goffredi, and R. C. Vrijenhoek, "*Osedax:* Bone-Eating Marine Worms with Dwarf Males," *Science* 305 (2004): 668–71.

2. K. Fulton-Bennett, "Whale Falls: Islands of Abundance and Diversity in the Deep Sea," *Ecosystem Observations* (2002), http://montereybay.noaa.gov/reports/ 2002/eco/mammals.html (accessed September 24, 2007).

3. S. K. Goffredi, C. K. Paull, K. Fulton-Bennett, L. A. Hurtado, and R. C. Vrijenhoek, "Unusual Benthic Fauna Associated with a Whale Fall in Monterey Canyon, California," *Deep-Sea Research* 1, no. 51 (2004): 1295–1306.

4. C. R. Smith and A. Baco, "Faunal Succession on Deep-Sea Whale Falls" (symposium presented at the Monterey Bay Aquarium Research Institute, Monterey, CA, March 27, 2002). Available at www.mbari.org/seminars/ 2002/winter2002/mar27_smith.html (accessed September 24, 2007).

18. Let's Talk about Sex, Baby

1. P. Clapham, "How Whales Do It," unpublished manuscript.

2. R. R Baker and M. A. Bellis, *Human Sperm Competition* (London: Chapman & Hall, 1995).

3. H. Whitehead, "Society and Culture of the Sperm Whale," *WGBH Forum Network,* November 10, 2004. Available at http://forum.wgbh.org/wgbh/ forum.php?lecture_id=1670 (accessed September 24, 2007).

4. J. C. D. Gordon, *Sperm Whales* (Grantown-on-Spey, Scotland: Colin Baxter, 1998), 22–25.

5. "Sperm Whales: The Real Moby Dick—Meet Jonathan Gordon," *Nature: PBS Online,* 1999. Available at www.pbs.org/wnet/nature/spermwhales/html/gordon.html (accessed September 24, 2007).

6. H. Yurk, L. Barrett-Lennard, J. K. B. Ford, and C. O. Matkin, "Cultural Transmission within Maternal Lineages: Vocal Clans in Resident Killer Whales in Southern Alaska," *Animal Behavior* 63 (2002): 1103–19.

7. J. Roughgarden, *Evolution's Rainbow: Diversity, Gender and Sexuality in Nature and People* (Berkeley: University of California Press, 2004).

8. G. Soeli, quoted in A. Doyle, "Homosexual Animals Exhibit Opens," *News.com.au,* October 12, 2006. Available at www.news.com.au/story/0,23599,20571062–1702,00.html (accessed September 24, 2007).

19. Missing Meat

1. The World Conservation Union, "Carcass of Critically Endangered Whale Found in Fishing Gear off Japan's Coast" (news release, February 1, 2007), www.iucn.org/en/news/archive/2007/02/01_pr_gray_whale.htm (accessed September 24, 2007).

2. Environment News Service, "Whales May Not Survive Sakhalin Oil Operations, Panel Finds," February 17, 2005, www.ens-newswire.com/ens/feb2005/2005–02–17–03.asp (accessed September 24, 2007).

3. D. W. Weller, A. M. Burdin, B. Wÿrsig, B. L. Taylor, and R. L. Brownell, "The Western Gray Whale: A Review of Past Exploitation, Current Status, and Potential Threats," *Journal of Cetacean Research and Management* 4 (2002): 7–12; D. W. Weller et al., "Gray Whales *(Eschrichtius robustus)* off Sakhalin Island, Russia: Seasonal and Annual Patterns of Occurrence," *Marine Mammal Science* 15 (1999): 1208–27.

4. D. W. Weller, Y. V. Ivashchenko, G. A. Tsidulko, A. M. Burdin, and R. L. Brownell, "Influence of Seismic Surveys on Western Gray Whales off Sakhalin Island" (working paper SC/54/BRG14, Cambridge: International Whaling Commission, 2002).

5. The World Conservation Union, "Independent Scientific Review Panel

Executive Summary: Impacts of Sakhalin II Phase 2 on Western North Pacific Gray Whales" (news release, February 16, 2005), www.iucn.org/ themes/business/Docs/ISRP_Executive_summary_in_English.pdf (accessed September 24, 2007).

6. World Wildlife Fund, "Shell Project May Doom Western Gray Whales to Extinction" (news release, February 16, 2005), www.worldwildlife.org/news/ displayPR.cfm?prID=183 (accessed September 24, 2007).

7. A. E. Kramer, "Russia Re-energized by Its Natural Resources," *International Herald Tribune,* April 9, 2006, www.iht.com/articles/2006/04/ 09/business/ rsibinvest.php (accessed September 24, 2007); J. Brooke, "Sakhalin Oil Treasure Is Finally Set to Flow," *International Herald Tribune,* September 28, 2005, www.iht.com/articles/2005/09/28/business/sakhalin.php (accessed September 24, 2007).

8. T. Hendricks, "On the Border," *San Francisco Chronicle,* December 10, 2005, www.sfgate.com/cgi-bin/article.cgi?file=/c/a/2005/12/ 10/MNG3OG620R1 .DTL (accessed September 24, 2007).

9. P. Jepson et al., "Gas-Bubble Lesions in Stranded Cetaceans," *Nature* (2003): 425, 575.

10. P. M. Vitousek, H. A. Mooney, J. Lubchenco, and J. M. Melillo, "Human Domination of Earth's Ecosystems," *Science* 277 (1997): 494–99.

11. I. Blühdorn, "Unsustainability as a Frame of Mind—and How We Disguise It: The Silent Counter-revolution and the Politics of Simulation," *Trumpeter* 18, no. 1 (2002): 1–11.

20. Shifting Scale

1. B. Henderson, "Open Letter to the Kansas School Board," May 2005, www .venganza.org/about/open-letter (accessed September 24, 2007).

2. J. Steinbeck, *The Log from the Sea of Cortez* (1951; repr., New York: Penguin Books, 1995), 51.

3. Ulrich Beck, "The Silence of Words and Political Dynamics in the World Risk Society" (symposium presented to the Russian Duma, translated from the German by Elena Mancini, November 2001). Available at http://logosonline .home.igc.org/beck.htm (accessed September 24, 2007).

4. R. Masters, "The First Eco-Writer: One Critic Claims That Steinbeck Was

the First of a New Breed of Environmental Writers," *Monterey County Weekly,* August 5, 2004.

5. To find out more about E. O. Wilson's 2007 TED prize wish, see "2007 TED Prize > E. O. Wilson," *TED: Ideas Worth Spreading,* www.ted.com/index.php/pages/view/id/166.

6. More details about the "Encyclopedia of Life" project are available at www.eol.org.

7. E. Douglas, "Darwin's Natural Heir," *Guardian,* February 17, 2001. Available at www.guardian.co.uk/Archive/Article/0,4273,4137503,00.html (accessed September 24, 2007).

8. Steinbeck, *Log from the Sea of Cortez,* 178–79.

DESIGNER: SANDY DROOKER

TEXT: 10/14 ADOBE GARAMOND

DISPLAY: AKZIDENZ GROTESK EXTENDED

COMPOSITOR: INTEGRATED COMPOSITION SYSTEMS

PRINTER: FRIESENS CORPORATION